THE PLATONIC SOLIDS BOOK

DAN RADIN

Copyright © 2008 by Daniel R. Radin

All rights reserved.

Published by CreateSpace Publishing

3-D renderings were created on a thirteen-year-old Macintosh computer using Infini-D software.

for my daughter Chelsea

TABLE OF CONTENTS

Introduction 1

Chapter 1   How to Make a Platonic Solid 3

Chapter 2   Dual Pairs 23

Chapter 3   Paper Polyhedra 31

Chapter 4   Origami 39

Chapter 5   Proof?! 51

Appendix   Lesson Plans 83

INTRODUCTION

The Platonic solids are the five most symmetric examples of polyhedra. The word, "polyhedra" is the plural form of the word, "polyhedron." I once read that the direct translation of the Greek word, "polyhedron," is "many seats." Apparently, "hedron" means seat and a "cat**hedra**l" is a place where people sit. Today, mathematicians generally translate it as a solid having many flat faces. But I can see how seats could work too. Maybe you can sit on them in many ways, or they contain many surfaces upon which they can sit. By "most symmetric," I mean that there are many ways that you can turn them around and have them still appear the same from different angles. I also mean that they look the same if you view them in a mirror.

The Platonic solids get their name from the Greek philosopher, Plato, who wrote about them. They were, in fact, known long before Plato by many different cultures. Plato wrote about them in his book *Timaeus*. This work was a mixture of philosophy, science, mathematics, and theology, which is not surprising since, at that time, the four fields were all considered part of the same whole. In *Timaeus*, Plato came up with an early version of atomic chemistry where all of matter was made up of combinations of these five shapes at a microscopic scale, and where it was these shapes that gave matter its properties.

It is not surprising that Plato believed that the gods had chosen these five most perfect forms from which to make all others. I know that I find them somehow intrinsically attractive, fascinating, and magical. I hope that you too will appreciate them as you read through this book and look at all of the beautiful images.

2

Chapter 1

How to Make a Platonic Solid

All five Platonic solids are made from three different regular polygons: the equilateral triangle, the square, and the regular pentagon. To be a Platonic solid, all of the polygon faces must be identical and the same number of faces must meet together at each vertex. "Vertex" is the word mathematicians use for the corners or points. The plural of "vertex" is "vertices."

The remainder of this chapter is devoted to instructions for making each of the five Platonic solids. After you finish reading it, you may want to skip to chapter three and start building your own models. Or, you may decide to just read on and enjoy the pretty pictures.

The Tetrahedron

The most basic Platonic solid is called the tetrahedron. The word "tetra" is Greek for four. The tetrahedron has four triangular faces. To make a tetrahedron, place three equilateral triangles point-to-point on a flat plane.

How to Make a Platonic Solid

With the center points still on the plane, swing the triangles up out of the plane.

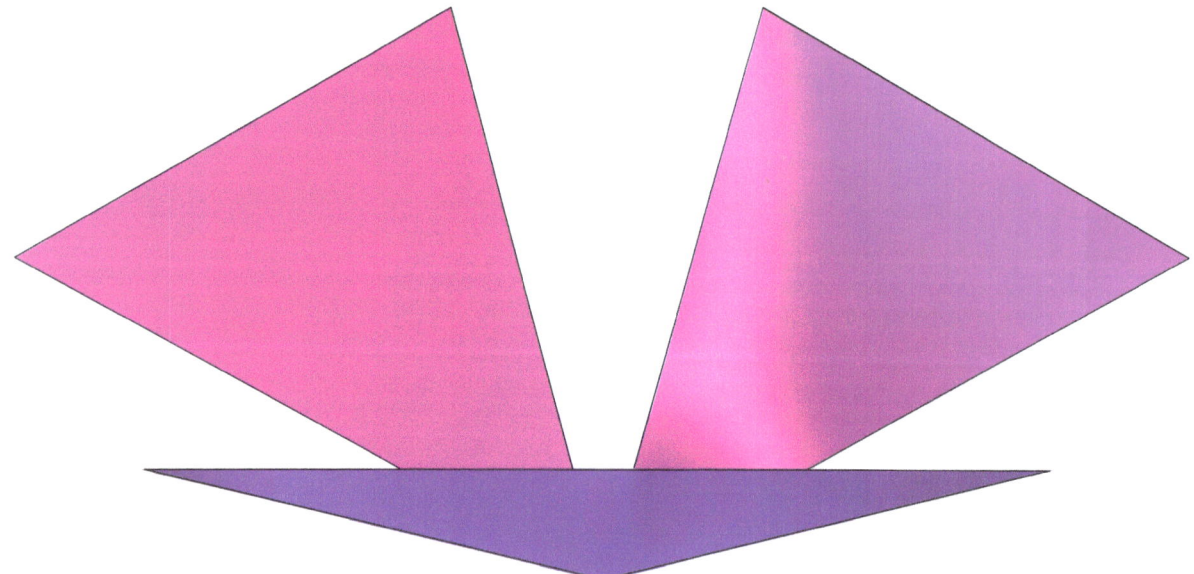

How to Make a Platonic Solid

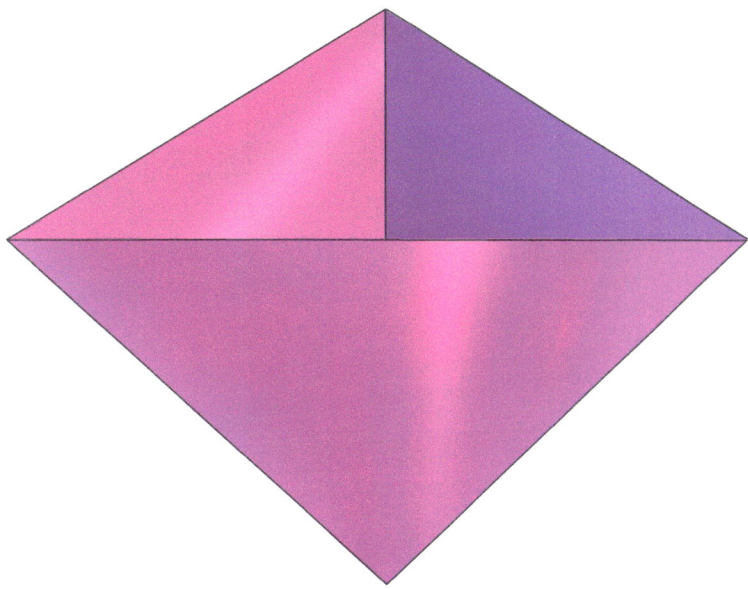

Place a fourth equilateral triangle on the top.

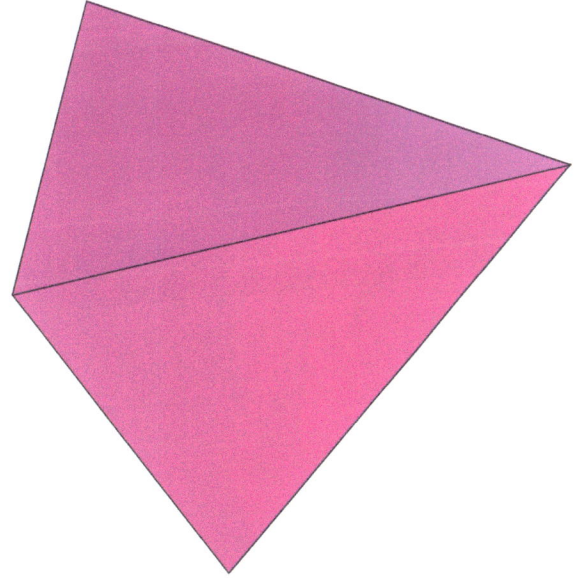

In a certain way, the tetrahedron is the most symmetric of the Platonic solids. Notice that each of its faces has three sides and that three faces meet at each vertex. The tetrahedron is the only Platonic solid with this dual nature between the number of sides on each face and the number of faces meeting at each vertex. You will see, in chapter two, that this feature of the tetrahedron has a very special consequence.

The Octahedron

The next triangle-based Platonic solid is called the octahedron. The word "octa" is Greek for eight. The octahedron consists of eight equilateral triangle faces. To make an octahedron, place four equilateral triangles point-to-point on a flat plane.

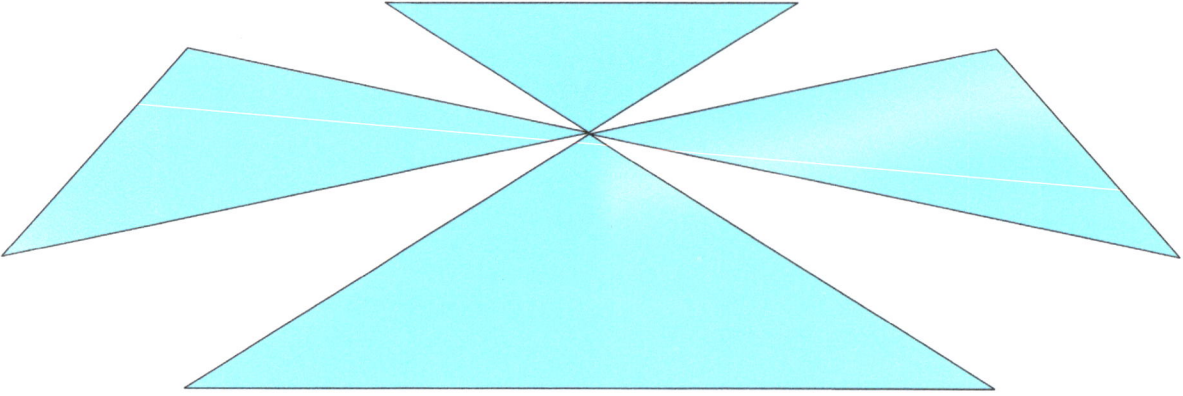

With the center points still on the plane, swing the triangles up out of the plane.

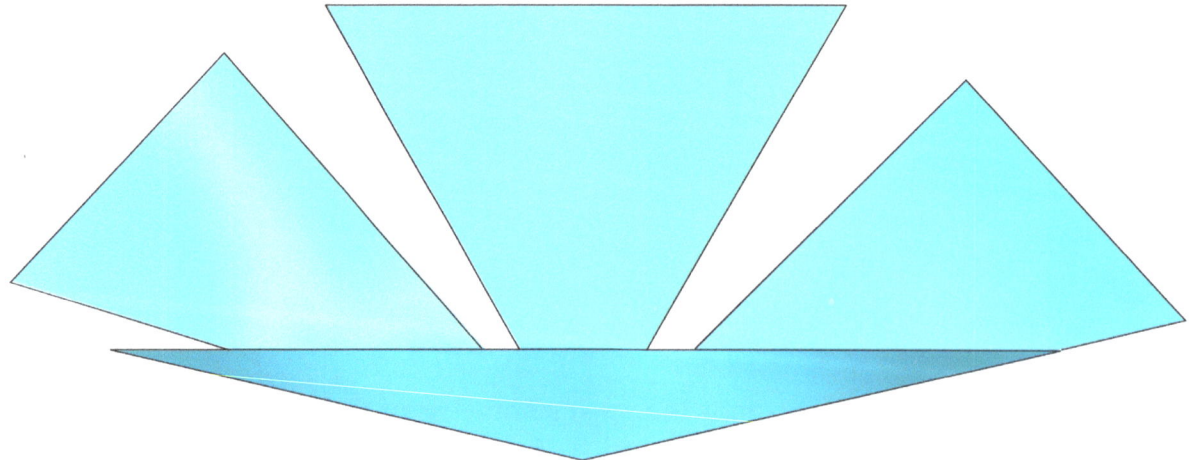

How to Make a Platonic Solid

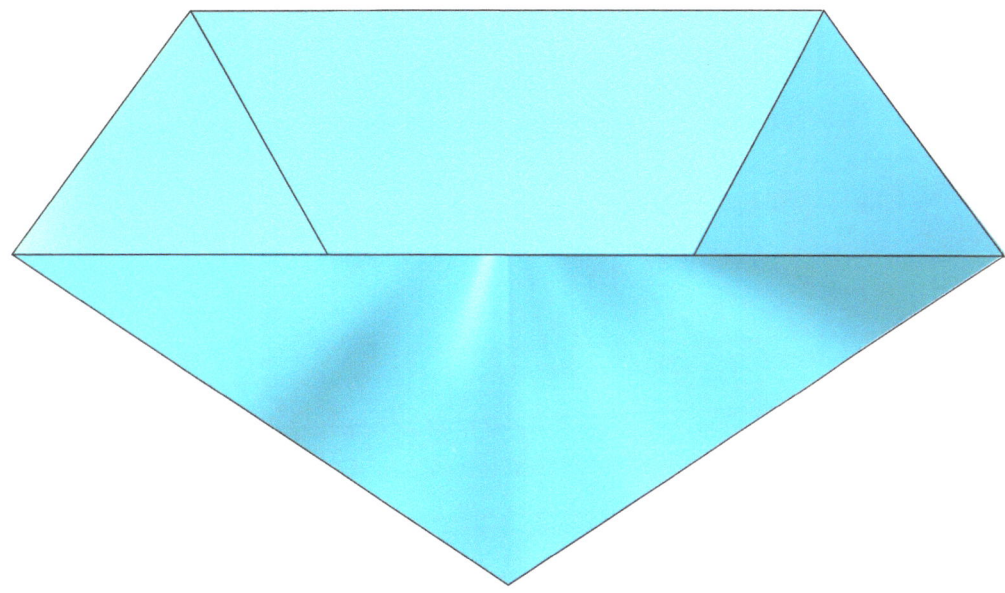

Start the whole process over again with four more triangles on the plane point-to-point. Swing the new triangles up out of the plane.

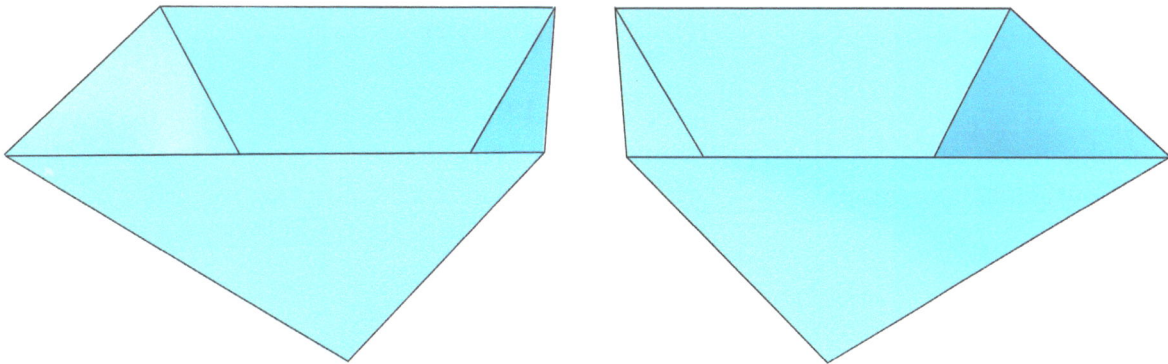

Place the second set of four triangles on top of the first set.

How to Make a Platonic Solid

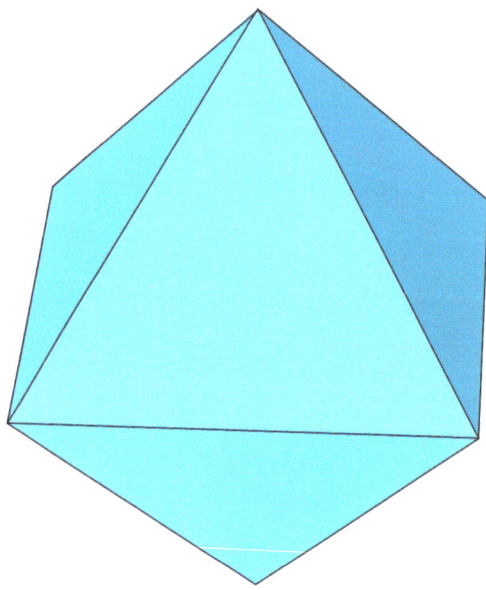

This is the octahedron, the second triangle-based Platonic solid.

The Icosahedron

The last triangle-based Platonic solid is called the icosahedron. The word "icosa" is Greek for twenty. The icosahedron consists of twenty equilateral triangle faces. To make an icosahedron, place five equilateral triangles point-to-point on a flat plane.

How to Make a Platonic Solid

With the center points still on the plane, swing the triangles up out of the plane.

10 How to Make a Platonic Solid

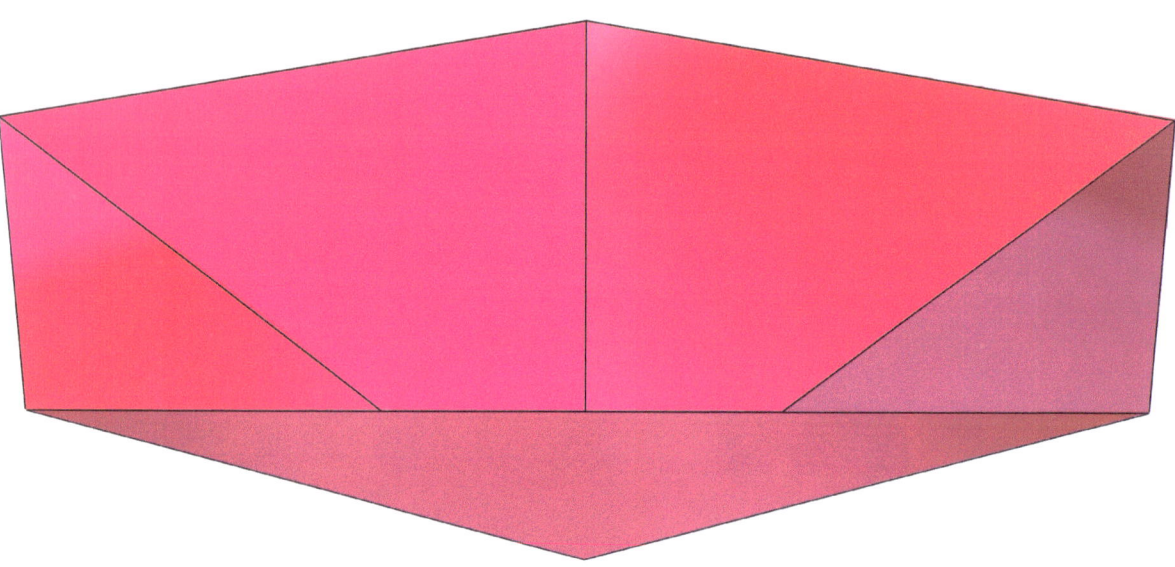

Now place five more triangles around the rim.

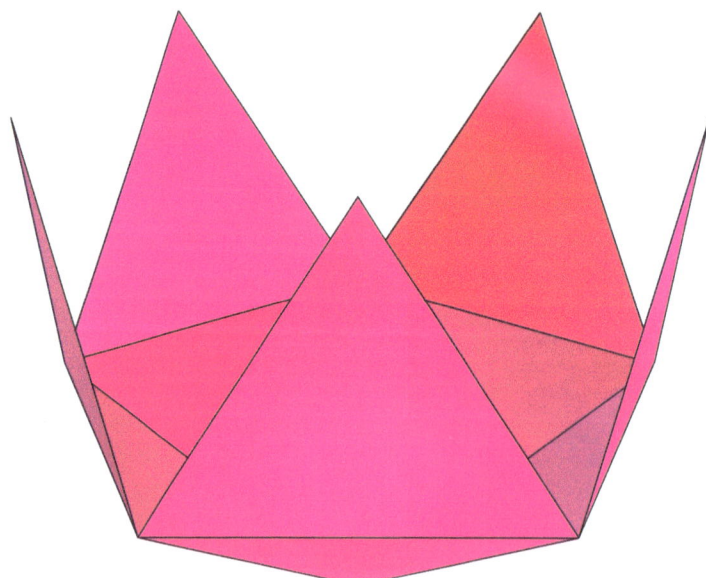

Now start the whole process over again with five more triangles on the plane point-to-point. Swing the new triangles up out of the plane, and place five more triangles around the rim.

How to Make a Platonic Solid

Place the second set of ten triangles on top of the first set.

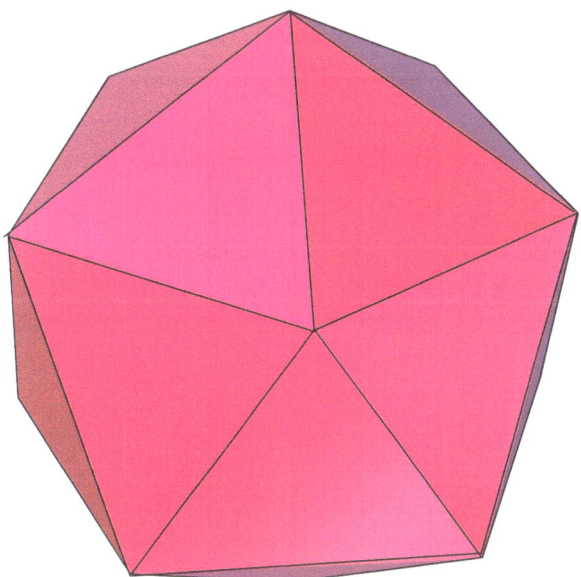

This is the icosahedron, the last of the triangle-based Platonic solids. This shape is the basis for most geodesic domes. The reason why there are no other triangle-based Platonic solids is that, when you place six equilateral triangles together on the plane point-to-point, they are already touching, and so you can't swing them up out of the plane.

So the only ways that equilateral triangles can meet at the vertices of a platonic solid are in sets of three (the tetrahedron), four (the octahedron), or five (the icosahedron).

The Cube

I am sure you are familiar with the Platonic solid that is made from squares. It is the familiar cube. It is also known as the hexahedron since "hexa" is Greek for six, and the cube has six faces. To make a cube, place three squares point-to-point on a flat plane.

How to Make a Platonic Solid

With the center points still on the plane, swing the squares up out of the plane.

14 How to Make a Platonic Solid

Start the whole process over again with three more squares on the plane point-to-point. Swing the new squares up out of the plane.

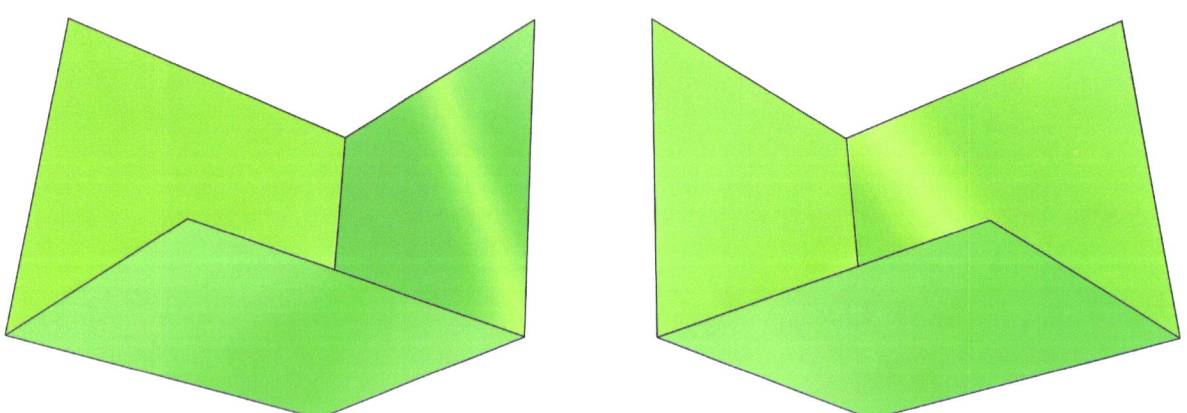

Place the second set of three squares on top of the first set.

How to Make a Platonic Solid   15

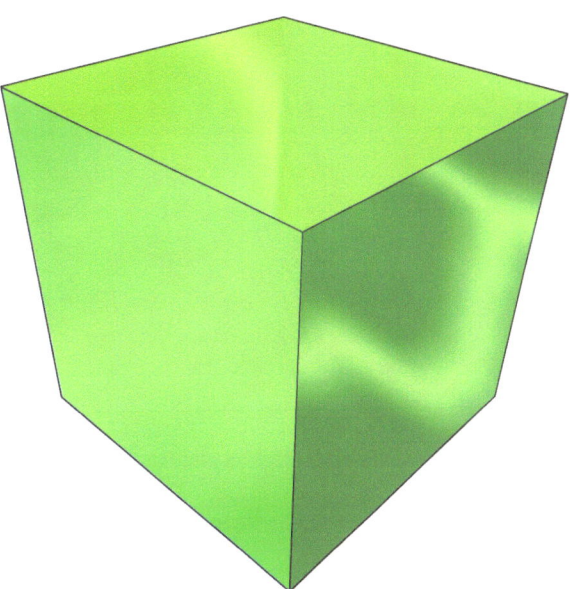

This is the cube or hexahedron, the only square-based Platonic solid. Perhaps you do not think the cube fits in with the other Platonic solids. It may seem too ordinary and perhaps not crystalline. Actually, there are several cube-shaped crystals. You will learn in the chapter on dual pairs that the cube fits in perfectly with the other Platonic solids.

The reason why there are no other square-based Platonic solids is that, when you place four squares together on the plane point-to-point, they are already touching, and so you can't swing them up out of the plane.

So the only way that squares can meet at the vertices of a platonic solid is in sets of three.

The Dodecahedron

The fifth and last Platonic solid is the dodecahedron. The word "dodeca" is Greek for twelve. The dodecahedron has twelve pentagon faces. To make a dodecahedron, place three pentagons point-to-point on a flat plane.

How to Make a Platonic Solid

With the center points still on the plane, swing the pentagons up out of the plane.

How to Make a Platonic Solid

Place three more pentagons around the rim.

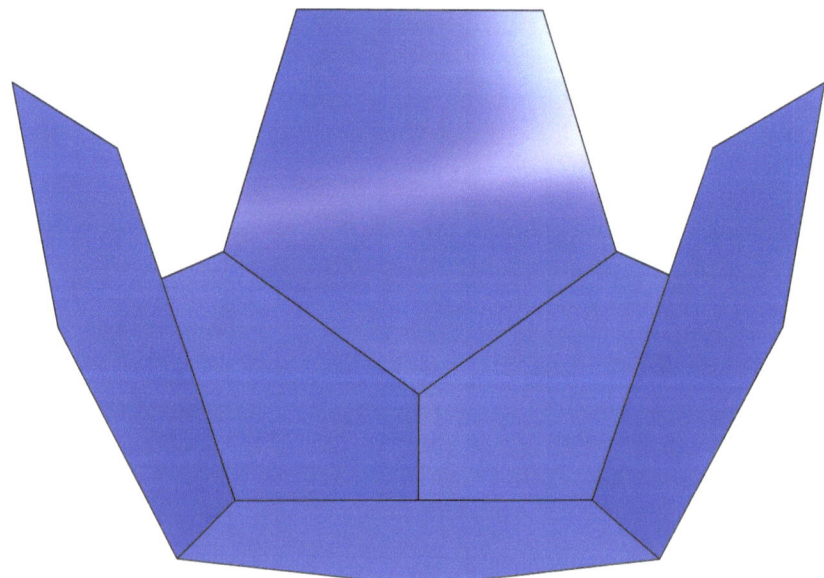

Start the whole process over again with three more pentagons on the plane point-to-point. Swing the new pentagons up out of the plane and place three more pentagons around the rim.

How to Make a Platonic Solid

19

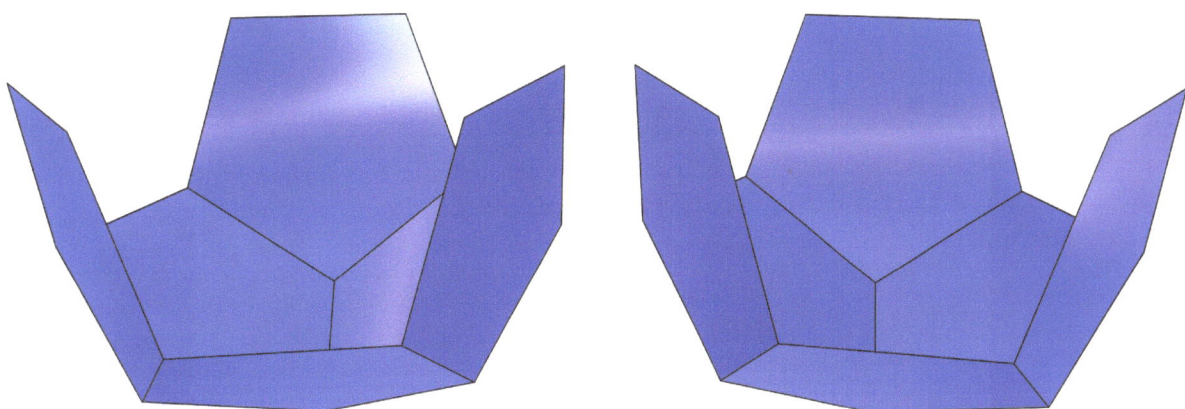

Place the second set of six pentagons on top of the first set.

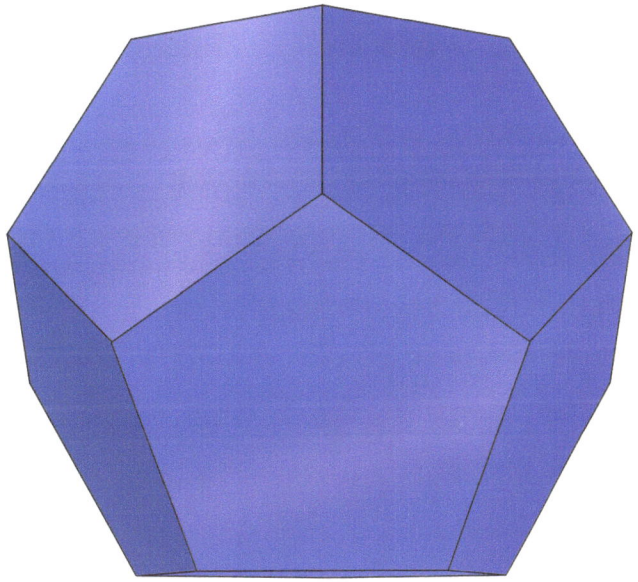

The dodecahedron is the basis of the pattern used for most soccer balls and volley balls.

The reason why there are no other pentagon-based Platonic solids is that you can't place more than three pentagons together on the plane point-to-point. They won't fit together.

How to Make a Platonic Solid

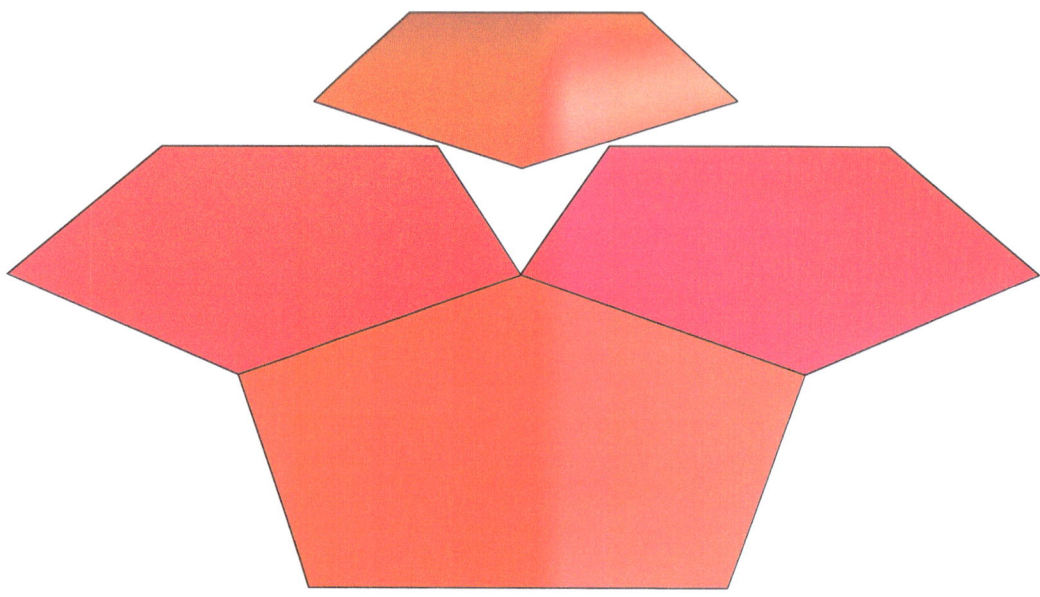

So the only way that pentagons can meet at the vertices of a Platonic solid is in sets of three.

Three hexagons fit together on the plane point-to-point with no room left to swing up.

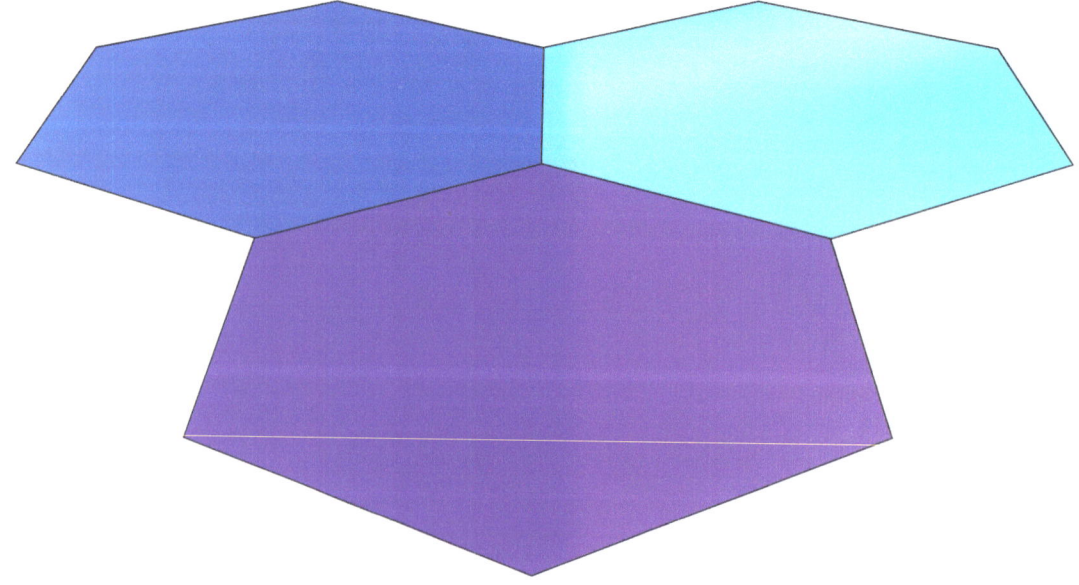

Three heptagons (seven-sided polygons) won't fit together on the plane point-to-point.

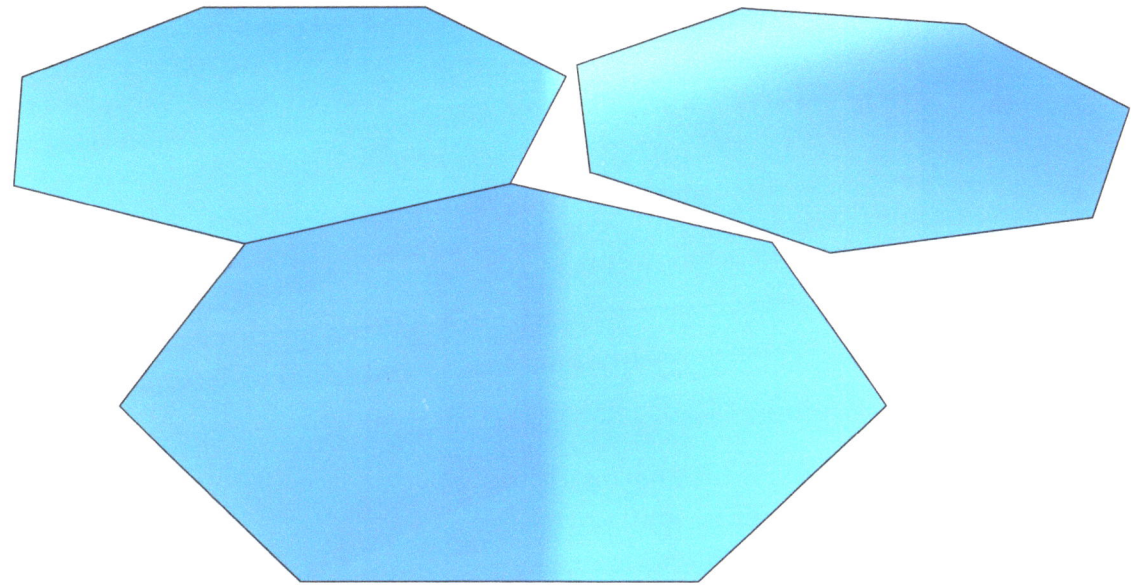

So the dodecahedron is the last of the Platonic solids. For some strange (some might say magical) reason, there are exactly five solids consisting of regular polygons, all of the same type, all fitting together in the same way. These are Plato's five sacred solids.

Questions to Ponder

- How could the preceding constructions be taught starting each solid with one regular polygon on the plane surrounded by other polygons placed together edge-to-edge?

- Why is it that, no matter what type and number of identical regular polygons you start with, if they can swing up out of the plane, they always fit together perfectly to complete a closed polyhedron?

- Why do crystals naturally form in the shapes of Platonic solids?

How to Make a Platonic Solid

- Why are people attracted to these forms even when they do not know about the mathematics behind them?

- How is it that our convention of there being 360 degrees in a full circle results in the fact that the first several regular polygons have integer degree measures in their angles, and how do these angles relate to the fact that there are exactly five Platonic solids?

- Given the number and type of polygons originally placed point-to-point on the plane, is there a way to calculate the total number of faces that will be in the completed Platonic solid?

- Is there a way to calculate the dihedral angle (the angle between two faces of the completed Platonic solid)?

- If one or more of the restrictions on Platonic solids is selectively removed or changed, what other classes of solids result?

Chapter 2

Dual Pairs

The Cube-Octa Pair

While the cube may not, at first, seem like a Platonic solid, it turns out that there is an octahedron hidden inside of every cube. If you connect the centers of the six faces of a cube, you will discover that they correspond to the six vertices of an octahedron in the interior of the cube.

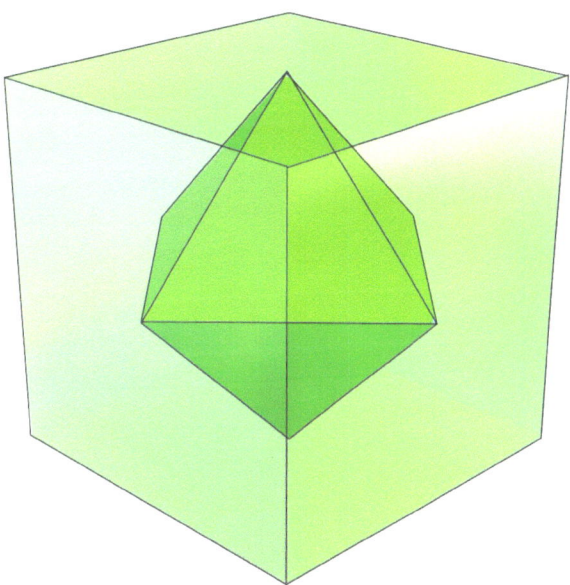

This relationship has to do with the numbers three, four, six, eight, and twelve. The cube has three faces meeting at each vertex, four sides to each face, six faces in all, and eight vertices in all. The octahedron has three sides to each face, four faces meeting at each vertex, six vertices in all, and eight faces in all. In this way, the faces of the cube are analogous to the vertices of the octahedron. Notice also, that they

 Dual Pairs

both have exactly twelve edges. The cube and the octahedron are a numerically bound pair.

cube		octahedron
3 faces meet at each vertex	←→	4 faces meet at each vertex
4 sides on each face	←→	3 sides on each face
6 faces in all	←→	8 faces in all
8 vertices in all	←→	6 vertices in all
12 edges in all	←→	12 edges in all

If you connect the centers of the eight faces of an octahedron, you will discover that they correspond to the eight vertices of a cube in the interior of the octahedron.

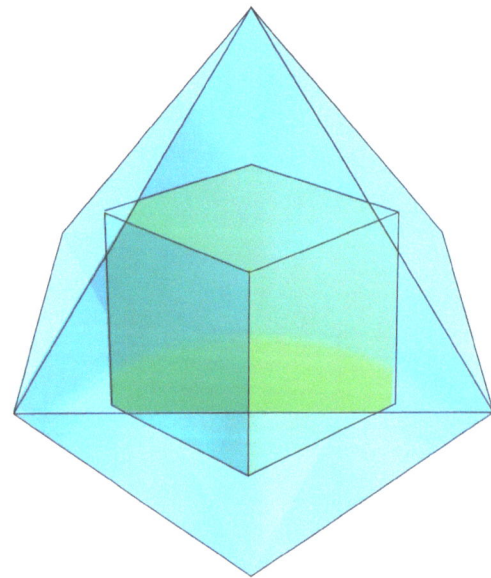

If you superimpose a cube with an octahedron of the same size, you get a beautiful compound solid.

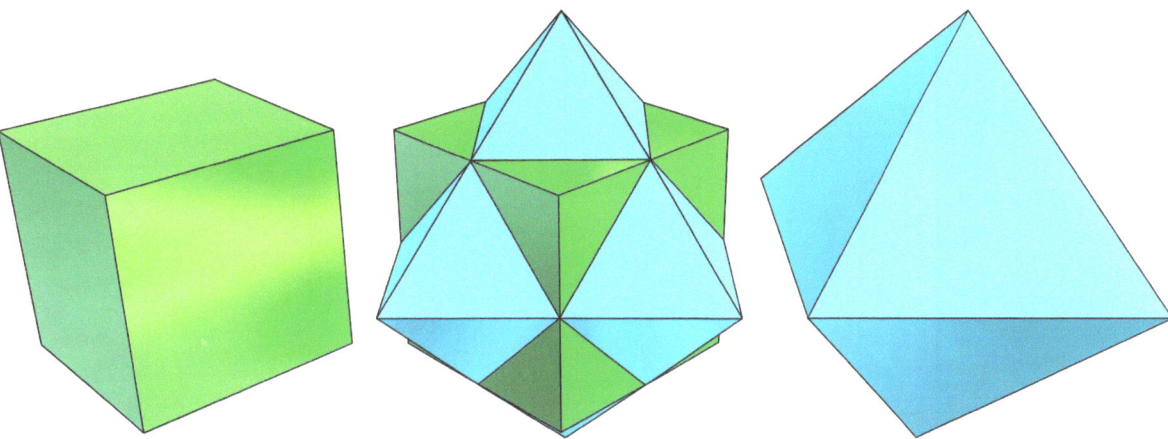

Notice that each square face has a square-based pyramid protruding from it, and that each triangle face has a triangle-based pyramid protruding from it. You can use the patterns in chapter three to build your own paper models of this and other compound solids.

The Dodeca-Icosa Pair

If you connect the centers of the twelve faces of a dodecahedron, you will discover that they correspond to the twelve vertices of an icosahedron in the interior of the dodecahedron.

Dual Pairs

This relationship has to do with the numbers three, five, twelve, twenty, and thirty. The dodecahedron has three faces meeting at each vertex, five sides to each face, twelve faces in all, and twenty vertices in all. The icosahedron has three sides to each face, five faces meeting at each vertex, twelve vertices in all, and twenty faces in all. In this way, the faces of the dodecahedron are analogous to the vertices of the icosahedron. Notice also, that they both have exactly thirty edges. The dodecahedron and the icosahedron are a numerically bound pair.

dodecahedron		icosahedron
3 faces meet at each vertex	⟷	5 faces meet at each vertex
5 sides on each face	⟷	3 sides on each face
12 faces in all	⟷	20 faces in all
20 vertices in all	⟷	12 vertices in all
30 edges in all	⟷	30 edges in all

If you connect the centers of the twenty faces of an icosahedron, you will discover that they correspond to the twenty vertices of a dodecahedron in the interior of the icosahedron.

Dual Pairs

27

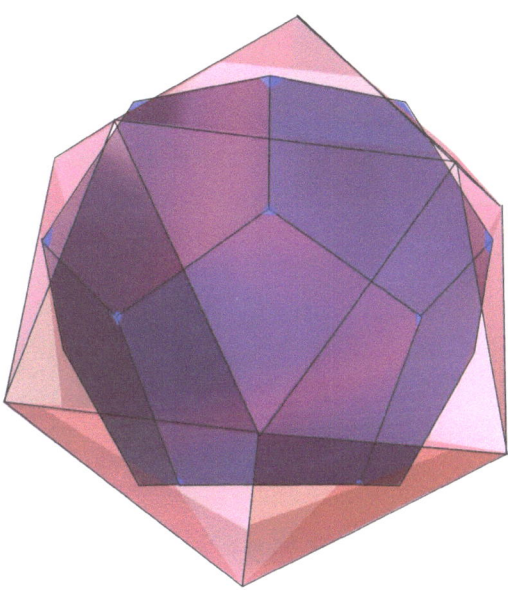

If you superimpose a dodecahedron with an icosahedron of the same size, you get a beautiful compound solid.

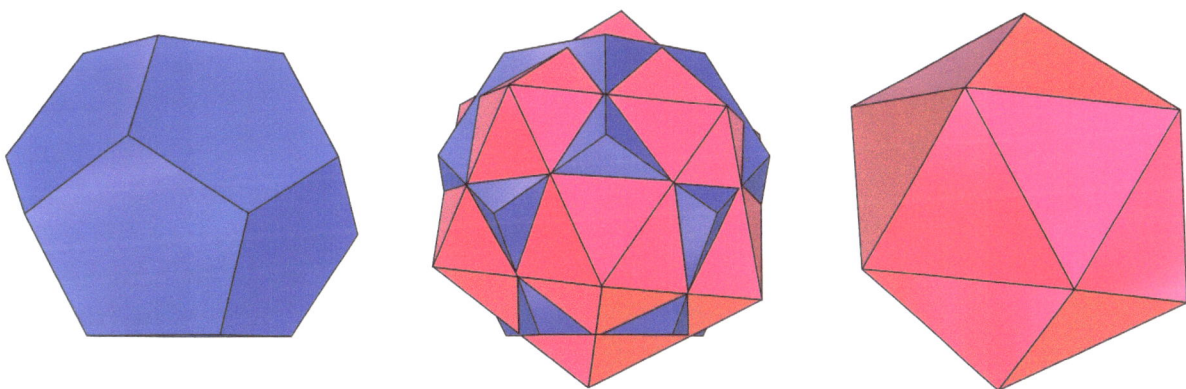

Notice that each pentagon face has a pentagon-based pyramid protruding from it, and that each triangle face has a triangle-based pyramid protruding from it. I find this to be a particularly fascinating solid. At one moment, I see the red icosahedron. Then, with the blink of an eye, I see the blue dodecahedron. Somehow though, my brain refuses to take in both shapes at once.

The Tetra-Tetra Pair

I have left this dual pair for last because it is perhaps the most surprising and intriguing. The tetrahedron is so symmetrical that it is its own dual. If you connect the centers of the four faces of a tetrahedron, you will discover that they correspond to the four vertices of a smaller tetrahedron in the interior of the original tetrahedron.

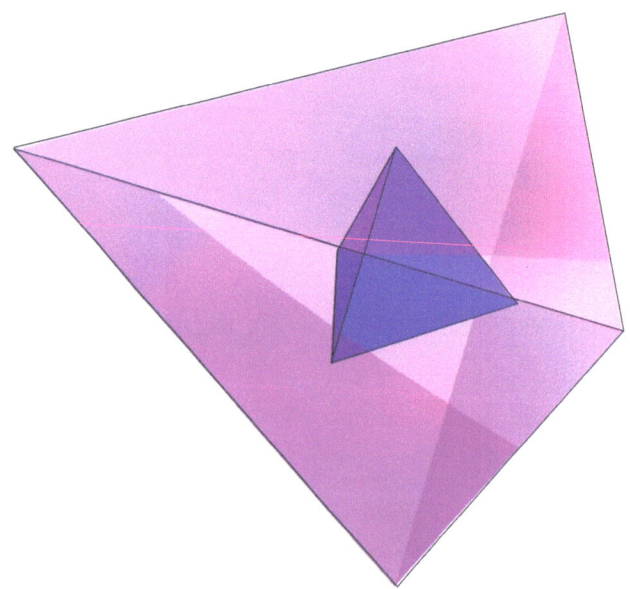

This relationship has to do with the numbers three, three, four, four, and six. The tetrahedron has three faces meeting at each vertex, three sides to each face, four faces in all, and four vertices in all. In this way, the faces of the tetrahedron are analogous to the vertices of the tetrahedron. Notice also, that both tetrahedra have exactly six edges. The tetrahedron and the tetrahedron are a numerically bound pair.

tetrahedron		tetrahedron
3 faces meet at each vertex	⤢	3 faces meet at each vertex
3 sides on each face	⤢	3 sides on each face
4 faces in all	⤢	4 faces in all
4 vertices in all	⤢	4 vertices in all
6 edges in all	↔	6 edges in all

Dual Pairs

If you superimpose a tetrahedron with a tetrahedron of the same size, you get a beautiful compound solid.

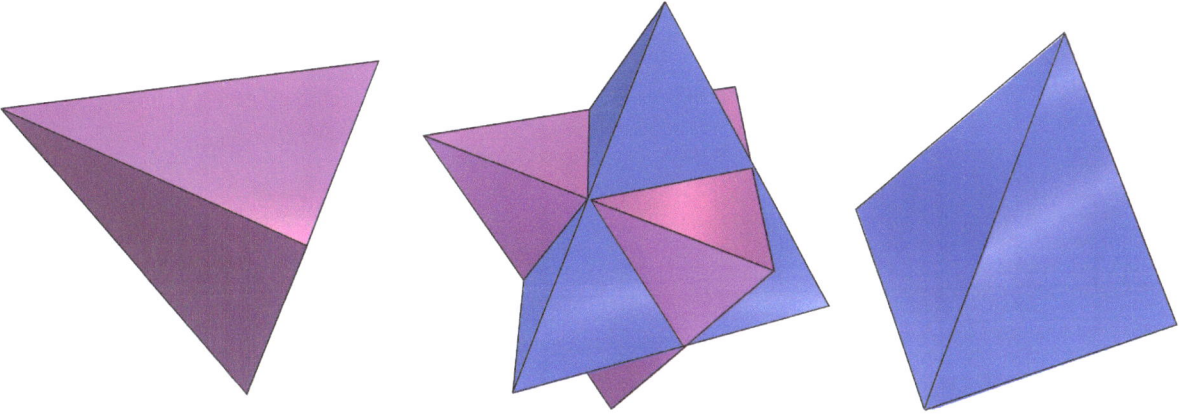

Notice that each triangle face has a triangle-based pyramid protruding from it. If you build a model of this particular compound solid, you will notice that it has a very strange quality. Even though the tetrahedron is arguably the most symmetrical of the Platonic solids, the compound solid of it and itself often appears asymmetrical. Viewed from certain angles, the two tetrahedra appear to be off center from each other.

I assure you that there is nothing wrong with the preceding image even though it looks somehow wrong. If you build your own model of this compound solid and hold it at certain angles, you will experience this optical illusion.

Questions to Ponder

- How are the dual pairs related in terms of planes of symmetry?

- Why is the compound solid of two tetrahedra the only one that can appear asymmetrical?

- For any given Platonic solid, what relationships exist among the number of faces meeting at each vertex, the number of sides on each face, the number of faces in all, the number of vertices in all, and the number of edges in all?

- Do any of the relationships referred to in the preceding question lead to the existence of dual pairs?

- If all of the pyramids are sliced off of a compound of two Platonic solid duals, what hidden interior solid is revealed?

Chapter 3

Paper Polyhedra

The following pages contain patterns for basic shapes. These patterns can be photocopied, cut out, and put together to make all kinds of polyhedra. For best results, photocopy the patterns onto colored card stock. Use a sharp point (such as a compass tip) and a straight edge to score along the dotted lines. Cut along the solid lines. (For the tetrahedron, cut along the light gray lines too.) Then fold along the dotted lines. The folded tabs can be attached together with glue, glue stick, staples, or double-sided clear tape. (If you use double-sided clear tape, apply the tape before cutting out the shapes.) The tabs can be folded to the outsides of the shapes, or they can be concealed on the inside. Use the stellation patterns to create pyramids on the faces of platonic solids, making them into compound solids. Use the small arrow heads to align the stellations. Have fun!!!

Triangle Faces

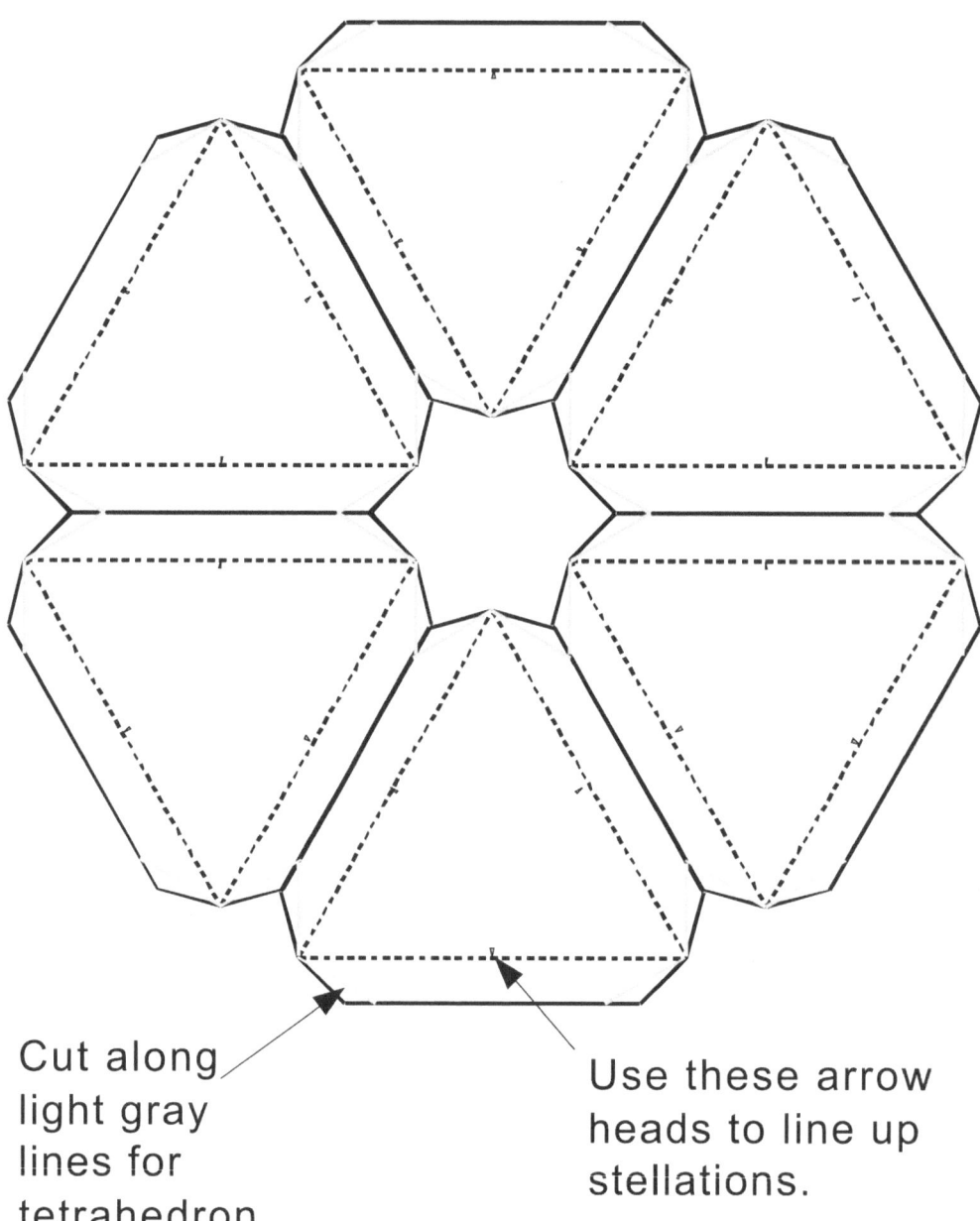

Cut along light gray lines for tetrahedron.

Use these arrow heads to line up stellations.

This pattern is from The Platonic Solids Book by Dan Radin.
This pattern is also available at www.PlatonicSolids.info.

Square Faces

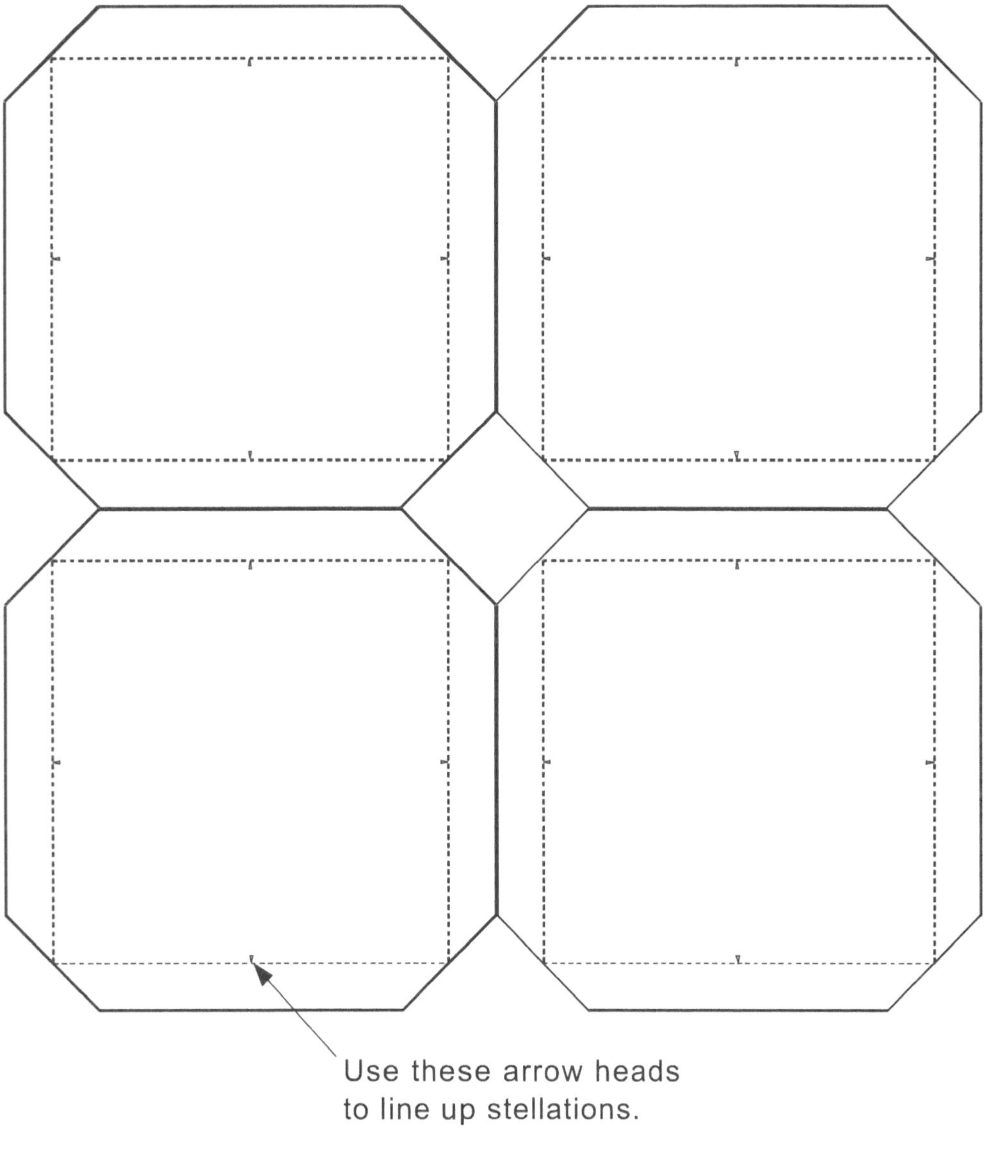

Use these arrow heads
to line up stellations.

This pattern is from The Platonic Solids Book by Dan Radin.
This pattern is also available at www.PlatonicSolids.info.

Paper Polyhedra

Pentagon Faces

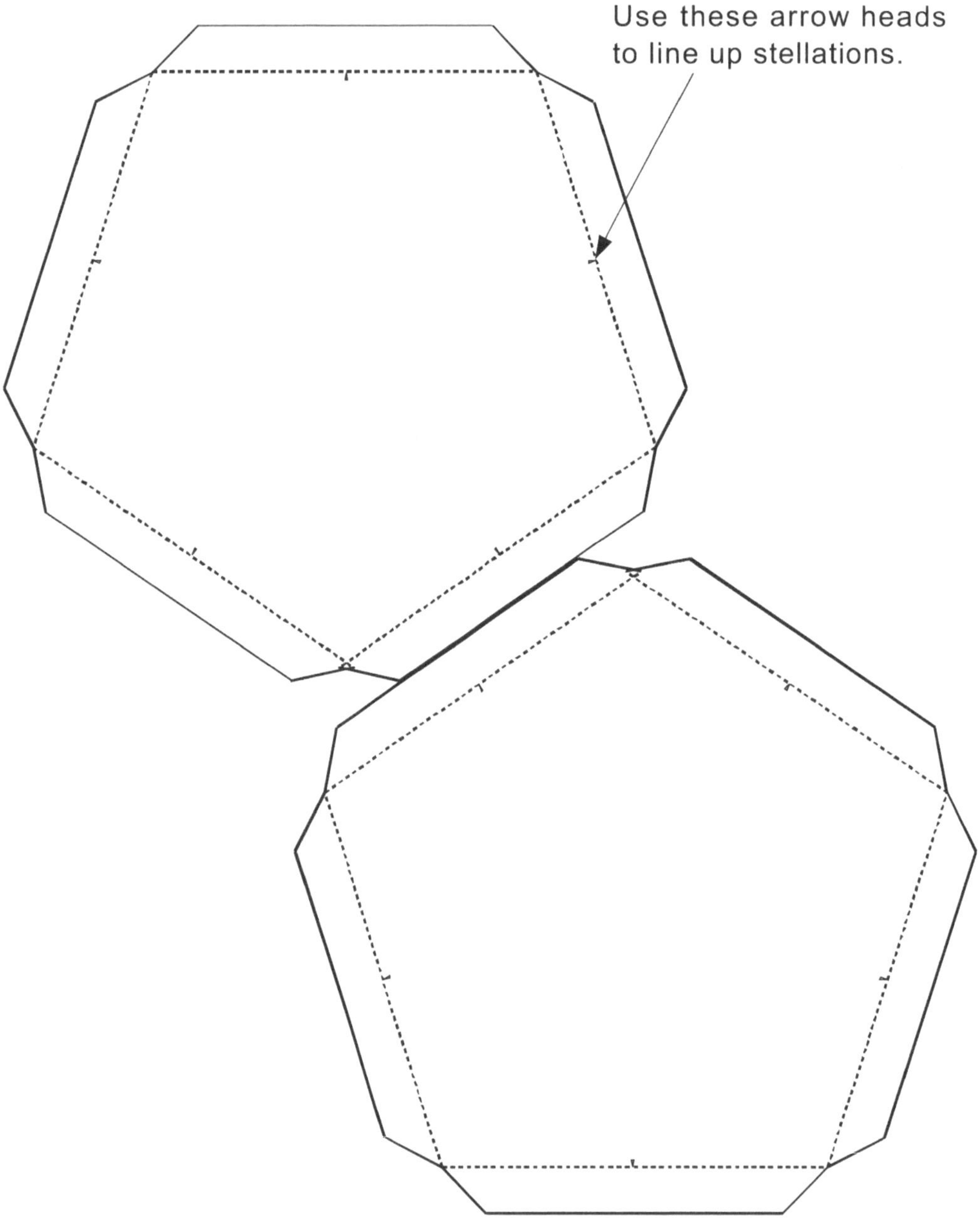

Use these arrow heads to line up stellations.

This pattern is from The Platonic Solids Book by Dan Radin.
This pattern is also available at www.PlatonicSolids.info.

Hexagon Face

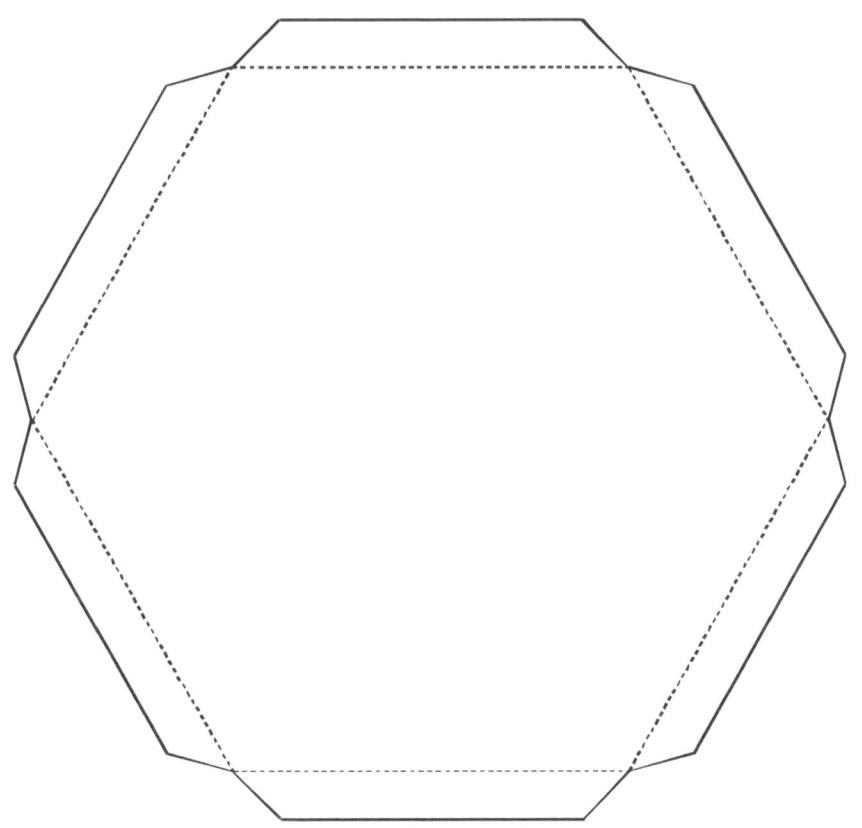

This pattern is from The Platonic Solids Book by Dan Radin.
This pattern is also available at www.PlatonicSolids.info.

Triangle-Based Pyramid Stellates

This pattern is from The Platonic Solids Book by Dan Radin.
This pattern is also available at www.PlatonicSolids.info.

Square-Based Pyramid Stellates

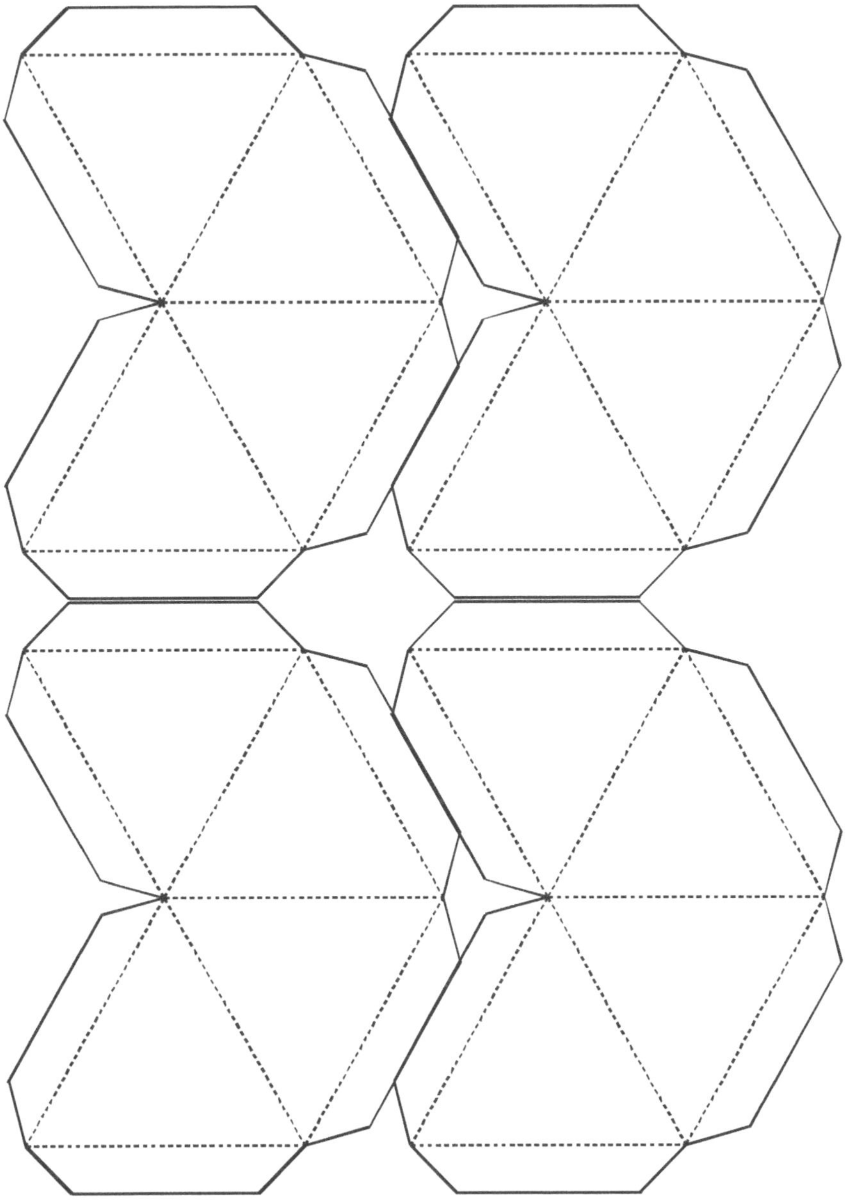

This pattern is from The Platonic Solids Book by Dan Radin.
This pattern is also available at www.PlatonicSolids.info.

Pentagon-Based Pyramid Stellates

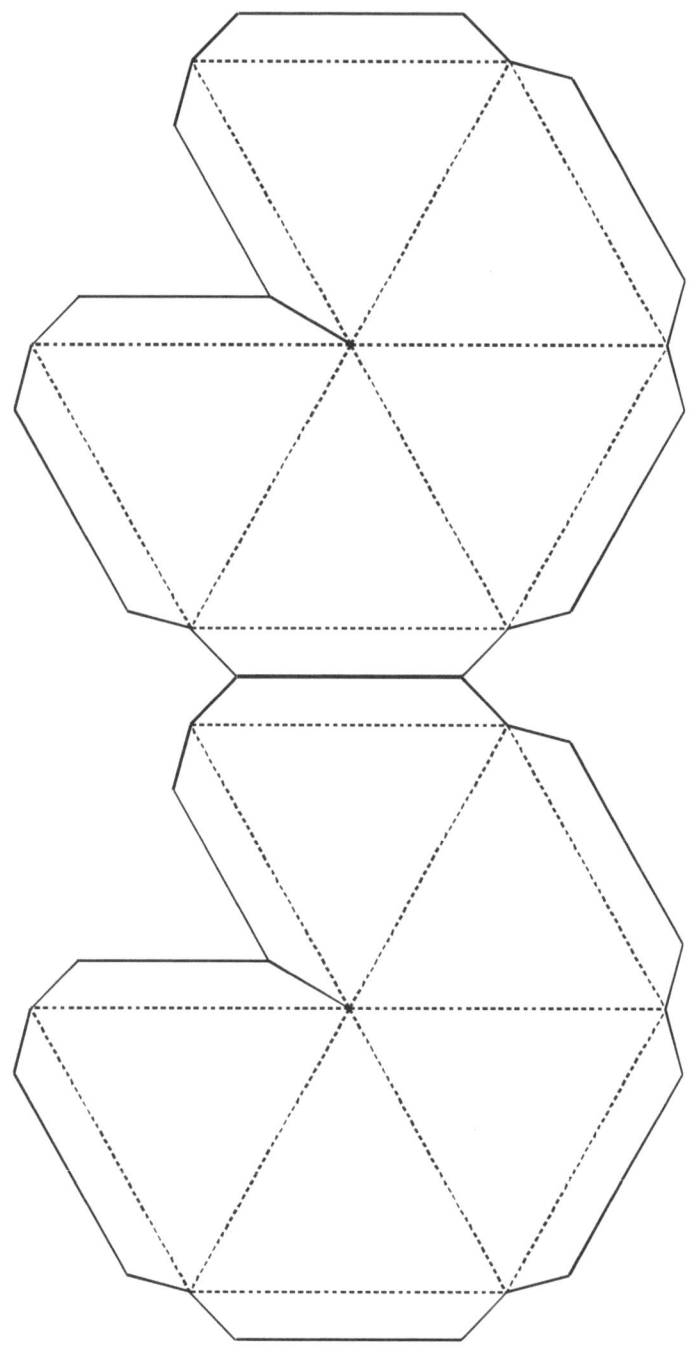

This pattern is from The Platonic Solids Book by Dan Radin.
This pattern is also available at www.PlatonicSolids.info.

Chapter 4

Origami

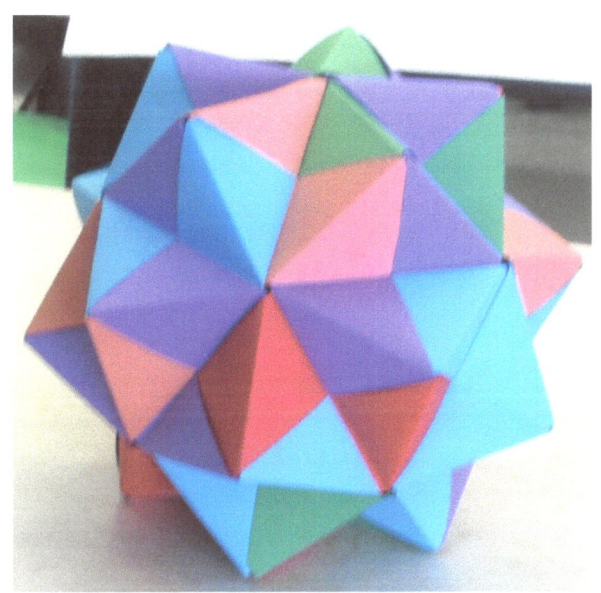

My friend Linda taught me how to make this wonderful origami polyhedron. It consists of twenty triangular pyramids arranged similarly to the faces of an icosahedron. It is made using standard origami techniques and classic square origami paper. The entire structure will hold itself together without any tape, glue, or staples. I find, however, that it is easier if I hide some tape on the inside. Otherwise, it is hard to keep it together before the last locking piece is inserted into place. My ninth grade students built all of the models shown and took all of the photographs. The pages that follow will guide you through the step-by-step process.

Step One

You will need 30 square pieces of paper. We used construction paper trimmed to square, but real origami paper will produce better results.

Step Two

If there is a nicer side of the paper, place the nicer side face down. Then fold the paper in half.

Step Three

Unfold the paper, and then fold the two sides in towards the center crease. The nicer side of the paper should be showing.

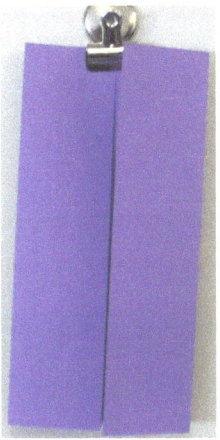

Step Four

Fold the paper in half perpendicular to the crease.

 ORIGAMI

Step Five

Turn the paper so that the open end faces down toward you.

Step Six

Fold the paper diagonally from the lower left corner to the upper right corner. Don't forget to fold the little triangle along with the large triangle.

Step Seven

Turn the paper over from left to right.

Step Eight

Again fold diagonally from the lower left corner to the upper right corner, including the little triangle.

Step Nine

Fold the lower right point up to the upper left corner. (Here's a hint: Don't fold it quite all the way to the corner. That way it will fit into its pocket better when you assemble the pieces.)

Step Ten

Turn the paper over and fold the other point upwards. (Once again, don't fold it all the way to the corner. This will make the wing fit into the pocket better later on.)

Step Eleven

The finished product should look like this. Repeat the preceding steps 29 more times so that you have 30 identical pieces. (Hint: Check to make sure they are all identical. If some are mirror images of others, you will run into trouble later on.)

Step Twelve

Slide the point of one piece into the pocket of another piece as shown.

 ORIGAMI

Step Thirteen

Interlock a third piece as shown to make a triangular pyramid with three wings coming off of it.

Step Fourteen

Start another pyramid coming off of one of the wings.

Step Fifteen

Eventually, you will have two interlinked pyramids as shown.

Step Sixteen

Start a third pyramid.

Step Seventeen

Now there are three complete interlocking triangular pyramids.

Step Eighteen

Keep going...

Step Nineteen

Keep going...

Step Twenty

When you get to the fifth pyramid, make them into a continuous ring of five. Then start working out from there. Always make sure the pyramids are in rings of five. Eventually, the whole thing will close up on itself and make a beautiful polyhedron. Theoretically, the entire structure should hold itself together without any tape or glue. But you can cheat by taping it on the inside where nobody will see once you are done.

CHAPTER 5

PROOF?!

This screenplay is adapted from *Proofs and Refutations: The Logic of Mathematical Discovery** by Imre Lakatos (November 9, 1922 – February 2, 1974). The original work, like my version, is in the form of a play taking place in a classroom. I wrote this adaptation of the first part of Lakatos' work for my community college students. I believe that this version, while far less thorough than Lakatos' original work, is easier for the layperson to read. If you find this chapter intriguing, I highly recommend the original work, which is truly a masterpiece on many levels. I am grateful for the late Mr. Lakatos' wonderful creation, and I want to reiterate that he deserves full credit for all of the ideas in this chapter.

Scene 1: The Proof

T: Good morning class! As you may remember, we spent last class studying polyhedra such as these models on my desk. We were trying to find a rule for relating the numbers of edges, vertices, and faces of polyhedra. As you know, for the polyhedra's simpler two-dimensional cousins, the polygons, there is a simple rule. In polygons, such as triangles, squares, pentagons, etc., the number of edges is equal to the number of vertices, and the number of faces is always one. We discovered last time that the rule for polyhedra seemed to be that the number of vertices plus the number of faces minus the number of edges was always two. Or $V + F - E = 2$. (*T* writes on the board explaining what each letter stands for.) By the way, this is known as Euler's Theorem of Polyhedra. But as of last class, no one had come up with a proof. So, as far as we know, our rule was really only a conjecture. Has anybody found a proof?

---

* This material is adapted from pp. 6-70, Proofs and Refutations: The Logic of Mathematical Discovery by Imre Lakatos edited by John Worrall and Elie Zahar Copyright © 1976 Cambridge University Press.

> **Euler's Theorem of Polyhedra**
>
> **V + F − E = 2**
>
> **V = number of vertices**
>
> **F = number of faces**
>
> **E = number of edges**

S: I haven't been able to come up with a proof but I'm satisfied that it's true. After all, we must have tried twenty different cases and it always worked. But if you have a proof, I'd like to see it.

T: Actually, I do have a proof, of sorts. It is in the form of a thought experiment. Let's use this rubber cube to demonstrate my proof.

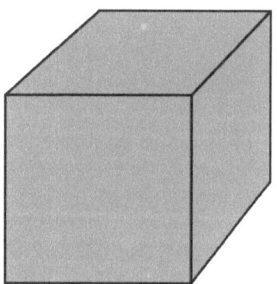

Of course you could use any polyhedron. I will show, in three steps, that the conjecture is true. I am going to prove, without counting, that for this cube, V + F − E = 2. For my first step, I will remove one face. (*T* cuts off one face with a knife.)

PROOF?!

Now, if and only if it was originally true that V + F - E = 2, then now V + F - E = 1, since I have certainly reduced F by one and hence reduced V + F - E by one from two to one. Can you see that V and E haven't changed? Now I staple the resulting defaced polyhedron onto a flat wooden board and mark the edges with white-out. (*T* proceeds to do this with a staple gun and some white-out.)

I contend that I could theoretically do this with any polyhedron. The resulting picture would be a map of the original polyhedron minus one face. For my second step, I draw diagonals in each polygonal face until the map is reduced to all triangles.

But you see that every time I add a diagonal, I increase the number of faces by one and I also increase the number of edges by one. These will cancel each other out and keep the expression, V + F - E, unchanged. Before my third step, I must first transfer the resulting map onto the board, since I will need to do some erasing. Now, I will begin removing triangles one by one. There are two ways to remove triangles. The first way involves removing one edge.

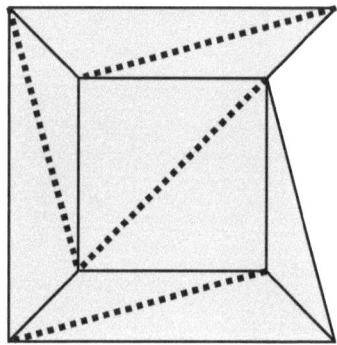

Thus we have lost one edge and one triangular face. Or E goes down by one and F goes down by one and you see, V + F - E doesn't change. The second way to remove a face is to remove two edges and a vertex like so.

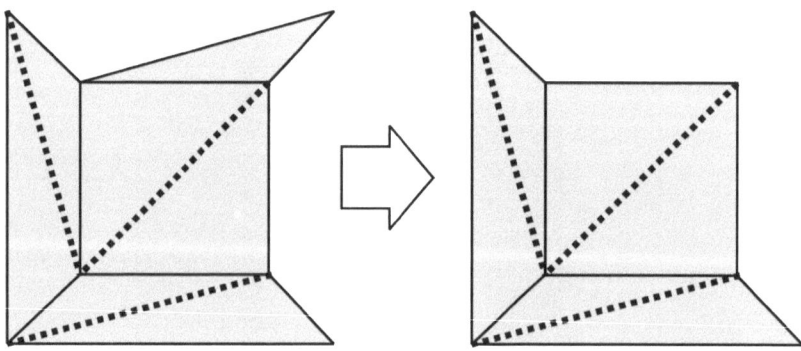

So E goes down by two, V goes down by one, and of course, F goes down by one, since we are removing a triangular face. Thus V + F − E remains unchanged. If we keep removing triangles, we will eventually end up with just one triangle. Now we know that for any triangle, V + F − E = 1. Since V + F − E remained unchanged throughout the triangle removal process, it must have equaled one ever since that top face was cut off of the cube. So you see, V + F − E had to have equaled two in the original polyhedron. Therefore, I have proven the conjecture that V + F − E = 2 for any polyhedron.

Scene 2: The Objections

D: Now that you've proven it, you don't have to call it a conjecture anymore. Now it's a theorem.

A: I'm not convinced. Are you sure that this works for any polyhedron? For instance, I wonder about your first step. Can any polyhedron be stretched flat on a board after the removal of one face?

B: Yeah, also, in that second step, I'm not convinced that every time you add a diagonal you get a new face.

G: And that third step, are you sure that there are only two ways that triangles can be removed? And are you sure that you will always end up with one triangle?

T: No, I'm not sure of any of these things.

A: So now we're worse off than before. Now we have three conjectures to prove instead of just one. How can you call what you just did a proof?

T: Well, if what you mean by a "proof" is something that establishes the truth of a conjecture, then I guess my thought experiment doesn't fit your definition.

D: Then what do you think a proof does?

T: That's a tricky question which I hope we will get to. For now, I propose to use the word, "proof," to describe a thought experiment that breaks a conjecture down into smaller sub-conjectures known as lemmas. By doing this, we have created a broader front by which to attack the original conjecture. Now we can look for counterexamples for any of the three sub-conjectures as a way of attacking the original conjecture.

> **Agenda**
>
> 1. What does a proof do?

Scene 3: Arguing the Third Lemma

G: As I said before, I suspect the third sub-conjecture or lemma as you call it. I suspect that there are other ways to remove triangles.

T: Suspicion is not a valid criticism.

G: What if I have a counterexample?

T: Conjectures ignore suspicion but they cannot ignore a counterexample.

G: Here's a counterexample. What if I remove a triangle from the inside of the network of triangles? Now I have removed a face without removing any edges or vertices. So I have changed $V + F - E$ and hence the third lemma is false! (G shows on board.)

T: You're right. But notice that while even a cube can be seen as a counterexample to my third lemma as you have shown, a cube is still not a counterexample to the original conjecture that $V + F - E = 2$. So you have a valid criticism of the proof, but not of the conjecture.

A: Does this mean that you will give up on this proof?

T: No, I will just improve the proof to stand up to the new criticism.

 PROOF?!

G: How?

T: First, let me explain a few new terms. There are two kinds of counterexamples: local counterexamples and global counterexamples. A local counterexample is an example that contradicts a lemma from within the proof, and a global counterexample is an example that contradicts the main conjecture that you're trying to prove. So you see, your counterexample is local but not global. It is a valid criticism of my proof but not of the original conjecture that V + F − E = 2.

G: So the conjecture may still be true, but I have shown that your proof does not prove it.

T: Yes, but I can easily fix my proof, or in particular, the lemma in question, so that your counterexample will no longer refute it. I only need to specify that the triangles must be removed from the boundary of the network, not from the inside. For instance, I could word it something like this: "Now remove the boundary triangles from the network one by one." So you see, it only took one small obvious adjustment to fix my proof.

G: I don't think it was such an obvious adjustment. It was actually pretty clever. Now I will show you that it was also false. If I remove boundary triangles as your new proof tells me I can, I can still run into trouble. What if I remove them in this order? (G shows at the board.) Now, as I remove this eighth triangle, I am removing two edges, one face, and no vertex, and hence V + F − E increases by one. If a boundary triangle is a triangle along the boundary, then you can't claim that this eighth triangle is not a boundary triangle.

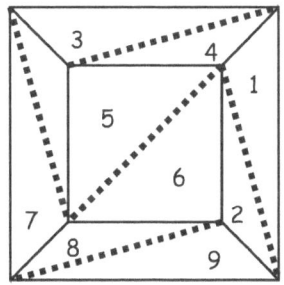

T: I could say that by "boundary" triangle I meant a triangle that does not disconnect the network, but honesty keeps me from doing this. Mathematical arguments starting with "I meant..." are rarely completely honest since part of the idea of a mathematical argument is to be perfectly clear about what you mean whenever you say anything. So yes, you got me. Here is a third version which will stand up to both of your counterexamples: "Remove the triangles one by one in such a way that V + F - E does not change."

K: Yes, your new lemma is certainly true. It says that if we remove triangles in such a way that V + F - E does not change, then V + F - E does not change. Big surprise there!

T: No, the lemma says that there is always an order for removing the triangles one by one without changing V + F - E until you get to the last triangle.

K: But how will we know what order to use and even if such an order exists? You started out with a thought experiment with definite instructions about removing triangles. Then you changed it to boundary triangles. Now you say to follow some particular order. But you don't say what the order is. How can we perform your experiment, even if it is a thought experiment, if you don't tell us exactly what to do? Your new lemma beats the counterexamples. But your thought experiment is no longer valid, and hence you no longer have a proof.

R: Actually, only the third step is gone.

K: You know, I'm not even sure I would call your new lemma an improvement. The first two versions looked true before we found the counterexamples. This last one is just a temporary patch. Do you really think this one will survive?

T: Single statements that look true are often quickly disproved. But more sophisticated statements that have been through several generations of criticism are more likely to actually end up being true.

O: What happens if this new more sophisticated lemma turns out to be false and you can't come up with a new patch?

T: Good question! Let's put it on the agenda for tomorrow.

 PROOF?!

> **Agenda**
>
> 1. What does a proof do?
>
> 2. What happens if we can't patch the lemma?

Scene 4: Counterexamples

A: I have a counterexample that refutes the first lemma. And it also refutes the main conjecture. That is, it can't be cut open and stretched flat, and it also doesn't satisfy V + F − E = 2. So it is a global counterexample and a local counterexample.

T: Great! Tell us what it is.

A: It is a solid defined as a cube with a cube-shaped hollow inside it like this. (A draws it on the board.) You can see that no matter what face you remove, it still won't stretch out into a flat map of edges and vertices. Also since for each cube, V + F − E = 2, my new shape must have V + F − E = 4. So it violates the first lemma and the original conjecture.

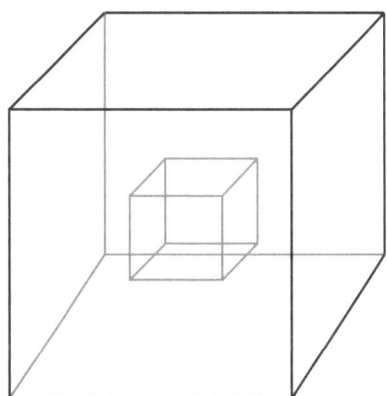

counterexample 1

T: Good job! We'll call it counterexample 1. (*T* labels it on the board.) Now what?

G: How can you take this so calmly? *A* has just wiped out your proof and the conjecture you're trying to prove. One counterexample is all it takes. It's time to scrap the whole conjecture.

T: I admit that the conjecture has taken a severe hit. I still believe that my proof, however, was successful. Remember that when I speak of a proof, I mean a way of breaking down a conjecture into a thought experiment containing several smaller conjectures called lemmas. These lemmas make it easier to analyze and challenge the original conjecture. Here, I think my proof was very successful. My proof certainly helped us learn more about the original conjecture.

A: So a local counterexample is a criticism of the proof, and does not hurt the conjecture, and a global counterexample hurts the conjecture, but does not invalidate the proof. But if a global counterexample knocks out the conjecture, what is left for the proof to prove?

G: Yeah, if the conjecture is gone, everything must go, including the proof!

D: But why do we have to accept this counterexample? The conjecture has been proven. It is now a theorem. It may not describe this so-called counterexample, but why should it give way? Let's get rid of the counterexample instead. I say this pair of nested cubes isn't a polyhedron at all. It's a monster created by *A*, and therefore it does not contradict the theorem.

A: Sure it's a polyhedron. A polyhedron is a solid bounded by polygonal faces.

T: Let's call this Definition 1. (*T* writes it on the board.)

 PROOF?!

> **Polyhedron**
>
> **Definition 1: a solid bounded by polygonal faces.**

D: Your definition is wrong. A polyhedron is a surface, not a solid. It has faces, edges, and vertices. And it can be stretched out on a board. The proper definition of a polyhedron is a surface consisting of a system of polygons.

T: Call it Definition 2. (*T* writes it on the board.)

> **Polyhedron**
>
> **Definition 1: a solid bounded by polygonal faces.**
>
> **Definition 2: a surface consisting of a system of polygons.**

D: So your so-called counterexample was really two polyhedra, one inside the other, i.e., two cubes. That would be like saying that a woman pregnant with a baby inside her is a counterexample to the conjecture that people have only one head each.

A: So my counterexample has made you come up with a new definition for polyhedron. Or is it your claim that you always meant a surface by saying polyhedron?

T: Let's just accept *D*'s new definition as Definition 2. Now that we mean a surface when we say polyhedron, can you still come up with a counterexample to refute the conjecture?

A: Actually, I can think of two. Take two tetrahedra that have a vertex in common or two tetrahedra that have an edge in common. (*A* draws them.) In each case, $V + F - E = 3$.

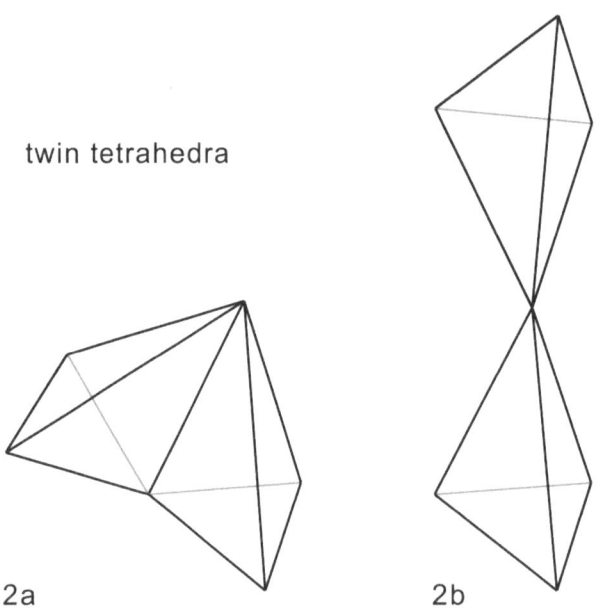

twin tetrahedra

2a   2b

T: We'll call them counterexamples 2a and 2b. (*T* labels them.)

D: Very imaginative! But of course, I didn't mean any system of polygons. I meant a system of polygons arranged so that exactly two polygons meet at any edge and there is a route inside the polyhedron from the inside surface of any polygon to the inside surface of any other polygon without crossing any edges or vertices. You can see how this knocks out your two sets of Siamese twin tetrahedra. (*D* demonstrates on the board.)

T: Call that Definition 3. (*T* writes it on the board.)

Polyhedron

Definition 1: a solid bounded by polygonal faces.

Definition 2: a surface consisting of a system of polygons.

Definition 3: a surface consisting of a system of polygons arranged so that exactly two polygons meet at any edge and there is a route inside the polyhedron from the inside surface of any polygon to the inside surface of any other polygon without crossing any edges or vertices.

A: You are the imaginative one, making up one definition after another to protect your pet theorem from my counterexamples. Why don't you just define a polyhedron as a system of polygons for which $V + F - E = 2$? This definition would settle the dispute forever!

T: Call that Definition P. (*T* writes it on the board.)

> **Polyhedron**
>
> **Definition 1:** a solid bounded by polygonal faces.
>
> **Definition 2:** a surface consisting of a system of polygons.
>
> **Definition 3:** a surface consisting of a system of polygons arranged so that exactly two polygons meet at any edge and there is a route inside the polyhedron from the inside surface of any polygon to the inside surface of any other polygon without crossing any edges or vertices.
>
> **Definition P:** a surface consisting of a system of polygons for which $V + F - E = 2$.

A: Then there would be no reason to study the subject any further.

D: I get your point but, on the other hand, any theorem can be contradicted by cleverly constructed monsters such as yours.

T: As you can see, when you come up with counterexamples to a conjecture, you often end up arguing about the definitions of the terms in the conjecture. In our conjecture, the ambiguity seems to be in the definition of the term, polyhedron. I made the mistake of assuming that we all agreed on what is and what is not a polyhedron. For now, let's not argue about which is the proper definition. Let's assume all of the definitions together. Does anybody have a counterexample that would work for even the most restrictive definition of a polyhedron?

K: Including Definition P?

T: No, not including Definition P.

G: I have one. I call it "the urchin." It's a star-polyhedron. Here's a model of it. (*G* displays model.) It consists of twelve star-pentagons like this one. (*G* shows model of a star-pentagon.) It has 12 vertices, 12 star-pentagonal faces, and 30 edges. You can count and see. So therefore V + F − E equals 12 + 12 − 30 which is −6, not 2. (*G* writes all of this on the board.)

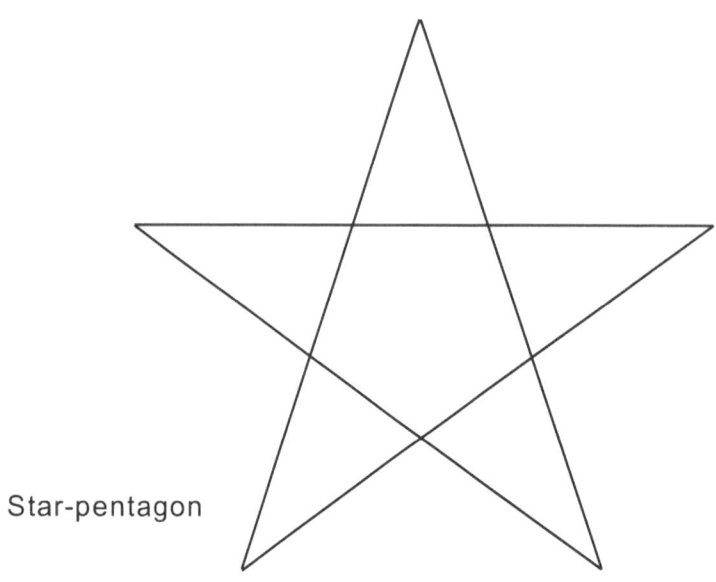

Star-pentagon

---

12 Vertices   12 Faces   30 Edges

V + F − E = 12 + 12 − 30 = **−6 not 2**

D: What makes you think your "urchin" is a polyhedron?

G: Don't you see? It's a polyhedron whose faces are star-polygons. So it consists of a system of polygons as required by definitions one and two. Exactly two polygons meet at any edge and it is possible to get from the inside of any polygon to the inside of any other polygon without crossing any edges or vertices. So you see, it satisfies Definition 3.

D: Then I guess you don't even know what a polygon is! A star-polygon is not a polygon. A polygon is a system of edges arranged so that exactly two edges meet at any vertex and the edges have no points in common except the vertices.

T: We'll call that Definition 1. (*T* writes it on the board.)

P<small>ROOF</small>?!

> **Polygon**
>
> **Definition 1:** a system of edges arranged so that exactly two edges meet at any vertex and the edges have no points in common except the vertices.

G: I agree that exactly two edges meet at any vertex. But I don't see why the edges have to have no points in common besides the vertices. I think the correct definition is just the first half of your definition.

T: We'll call that Definition 1'. (*T* writes it on the board.)

> **Polygon**
>
> **Definition 1:** a system of edges arranged so that exactly two edges meet at any vertex and the edges have no points in common except the vertices.
>
> **Definition 1':** a system of edges arranged so that exactly two edges meet at any vertex.

G: Look, if I just lift the edge of this model of a star-pentagon, it still satisfies the entire Definition 1 anyway. Now the edges have no points in common except the vertices. Your problem is that you limit yourself to the plane. You should let your polygons stretch out into space.

D: Can you tell me what is the area of your star-polygon, or do you claim that polygons don't have areas?

G: You were the one who claimed that polyhedra were not solids, that they were just surfaces. In that case polygons must be just closed curves consisting of edges and vertices, and not the area they enclose.

T: Let's save this debate for another time and get back to the task at hand. Does anyone have a counterexample that works for even the new definitions: 1 and 1'?

A: I have one. Look at this picture frame. (*A* shows the frame.) It passes every definition of a polyhedron. But if you count its edges, faces, and vertices, you will see that you get V + F - E = 0.

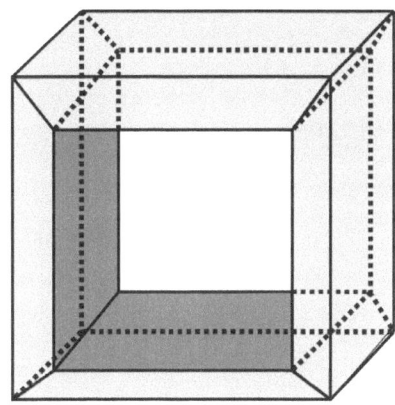

counterexample 4 (picture frame)

T: We'll label it counterexample 4. (*T* puts a label on it.)

B: I guess that does it. We've just been wasting our time. The conjecture is false.

A: Well *D*, aren't you going to say anything? Can't you define this new counterexample out of existence? Have you given up? Do you admit that we have finally shown the existence of non-Eulerian polyhedra?

Scene 5: Order versus Disorder

D: *A*, I am getting tired of your little non-Eulerian pests. You should really find another name for them though, instead of insisting on calling them polyhedra. I see Euler's theorem as a beautiful example of the order and harmony in mathematics. You seem to prefer to seek anarchy and chaos in mathematics. I don't see how we can ever resolve our differences.

A: Well you will say anything to preserve your precious order and harmony from supposed anarchists like myself.

T: Do we have a new definition to rescue the conjecture?

A: You mean the latest contraction of the concept of polyhedron. *D* simply sidesteps real problems with new definitions. He doesn't solve them.

D: I'm not contracting the concept. You're expanding it. For instance, your picture frame is obviously not a polyhedron.

A: Why not?

D: If you cut your picture frame with a plane like so, (*D* cuts frame on table saw.) you see that it has two completely disconnected polygon cross-sections. You will find that this is true for any plane passing through the inside of the frame.

A: Your point being…?

D: For a genuine polyhedron, there is at least one plane through any point such that the intersection of the plane with the polyhedron consists of a single polygon.

T: We'll call that Definition 4. (*T* writes it on board.)

> **Polyhedron**
>
> **Definition 1:** a solid bounded by polygonal faces.
>
> **Definition 2:** a surface consisting of a system of polygons.
>
> **Definition 3:** a surface consisting of a system of polygons arranged so that exactly two polygons meet at any edge and there is a route inside the polyhedron from the inside surface of any polygon to the inside surface of any other polygon without crossing any edges or vertices.
>
> **Definition P:** a surface consisting of a system of polygons for which $V + F - E = 2$.
>
> **Definition 4:** a surface consisting of a system of polygons arranged so that exactly two polygons meet at any edge and there is a route inside the polyhedron from the inside surface of any polygon to the inside surface of any other polygon without crossing any edges or vertices and there is at least one plane through any point such that the intersection of the plane with the polyhedron consists of a single polygon.

A: For each counterexample, you have a new definition which you claim to be just a deeper insight into the original concept of polyhedron. You have turned Euler's original beautiful conjecture that $V + F - E = 2$ into some sort of holy dogma to be followed blindly. (*A* leaves the room.)

 Proof?!

D: I can't understand why an intelligent mathematician like *A* wastes her(his) time with these monsters (s)he calls polyhedra. Monstrosities never serve any purpose either in nature or in the world of thought. Nature always follows a harmonious orderly pattern.

G: Biologists would argue with that. The theory of evolution is driven by mutations sometimes referred to as "hopeful monsters." I think *A*'s counterexamples are "hopeful monsters."

D: Well, *A* is gone and there will be no more monsters of any sort.

G: I've got a counterexample. It satisfies every definition we have, and yet V + F - E = 1. My counterexample is a cylinder. (*G* provides cylinder.) It has three faces: the top, the bottom, and the side, two edges: the circles, and no vertices.

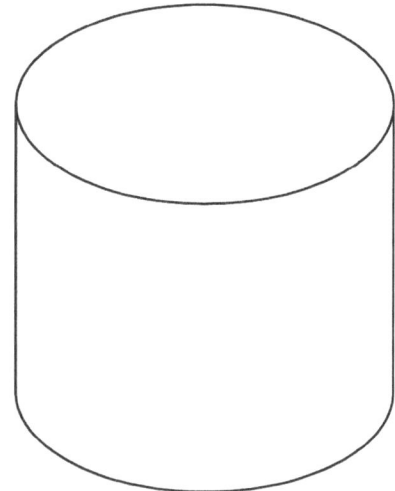

D: *A* stretched concepts, but you tear them. An edge has to have two vertices.

T: Is this another definition?

G: Why exclude edges with one or even zero vertices? You used to contract concepts. Now you practically define them out of existence.

D: I don't see any purpose for your so called counterexamples. You're trying to include so many weird monsters into the term, "polyhedron," that there is barely any room for ordinary named polyhedra.

G: I believe that the only way to gain a deep understanding into a concept is to study it at the edges where things get more real and interesting. If we want to know about what a polyhedron is, we must look at the edge, the lunatic fringe, of polyhedra.

T: I must agree that while D has done an excellent job of defending the theorem against monstrous counterexamples, his method is perhaps not the most useful. We need to somehow treat these monsters with more respect. D is still holding to the idea that a proof must prove what it sets out to prove. I see a proof as the decomposition of a conjecture into subconjectures. Hence, even if the conjecture is false, the proof can still be interesting.

Scene 6: Exceptions

B: That was very confusing. But before you explain it further, there is something I have to say.

T: Yes, let's hear it.

(*A re-enters.*)

B: I don't think we should call these things monsters or even counterexamples. This just sets them up as the enemy. I think they are natural and I propose that we call them exceptions. They help to restrict the domain of the conjecture.

S: I agree with you. I think there are three kinds of conjectures: those that are always true, those that are false, and those that are generally true with certain exceptions.

E: What are you talking about?

S: There is a difference between a false conjecture and one that is subject to restrictions. After all, they say "the exception proves the rule."

E: (to K) (S)he needs to learn something about logic.

D: I am embarrassed to say that I think A and I are on the same side in this discussion. At least we both agree that the conjecture must be either true or false. S's idea of a third category: true but subject to exceptions, makes no sense. You can't have a mathematical theorem that has exceptions. It's too muddy for mathematics. It's starting to sound like English grammar.

A: I agree.

E: I must side with D's original arguments against the monsters. It's not so much a matter of protecting the theorem from the likes of A. My interest is in protecting mathematics from S's kind. I don't think we need exceptions if we are careful with our definitions.

A: You could just as easily side with my counterexamples and label the conjecture false.

E: It makes more sense to me to reject your monsters than to reject a perfectly good proof.

T: Let's go back to B's and S's idea of renaming the counterexamples as exceptions.

Scene 7: Refining the Domain

B: Actually, I don't agree with S's idea of a third category vaguely named, "true with exceptions." I see the exceptions as useful tools, not for attacking the conjecture, but for refining it and narrowing it down until we know exactly where it does and does not apply.

T: So what would be the precise domain of Eulerean polyhedra?

B: All polyhedra that have no cavities (like the nested cubes) and no tunnels (like the picture frame).

T: Are you sure?

B: Yes, I'm sure.

T: What about the twin tetrahedron? (*T* points them out.)

B: Okay, no cavities, no tunnels, and no multiple structures.

T: I like your idea, but I wonder if it is really now perfect and unambiguous. How do you know it excludes all exceptions?

B: Can you name one I don't exclude?

A: What about my urchin? (*A* points to it.)

G: What about my cylinder? (*G* points to it.)

T: Even if we don't have an actual exception to show you, how can you be sure none exist?

B: I guess you're right. It was ridiculous to think we could generalize from our small study of regular polyhedra. I'm surprised we didn't find more exceptions. I think we can safely restrict our domain to convex polyhedra though. It was the concave ones that gave us trouble.

G: What about my cylinder? It's convex.

B: It's a joke.

T: Setting aside the question of the cylinder, I still have some criticism of your method. You have retreated for safety to include just convex polyhedra. But perhaps you have gone too far and excluded many fine Eulerean polyhedra. The original conjecture may have been an overstatement. But yours sounds like an understatement. At the same time, how do you know that there are no convex exceptions? Perhaps yours is also an overstatement. My other

B: problem is that yours sounds like a guess. Where is the proof? Are you saying we no longer need a proof?

B: I didn't say that.

T: That's true, but you did discover that the proof didn't prove the original conjecture. Does it prove your new conjecture?

B: Well…

E: Here we see that we must not abandon the theorem just because of a few monsters disguised as exceptions.

B: Actually, I reject the original conjecture and its supposed proof. Both have exceptions. But I will restrict both the conjecture and the proof to a proper domain, thereby creating a true and rigorous theorem and proof. For instance, not all polyhedra can be stretched flat on a plane after having one face removed, but all convex polyhedra can be. So the proper theorem is that all convex polyhedra are Eulerian. And it can be rigorously shown that each lemma of the proof holds up under the restricted domain.

T: How do you know that this convex polyhedron restriction isn't just guesswork like the tunnel and cavity guesses?

B: This time it is insight, not guesswork!

T: I admire a guess because it shows courage and modesty. Insight, I question.

B: Question all you like, but do you have a counterexample to my theorem that all convex polyhedra are Eulerian?

T: Certainly, you have no way of being sure that I don't. There is no proof in your method.

B: Do you have the perfect method?

T: No, but I think I can at least show you a method that incorporates counterexamples and proof.

B: I'm all ears.

R: May I get a word in here?

T: Go ahead.

Scene 8: Refining the Interpretations

R: I reject *D*'s method of disqualifying supposed monsters. I also reject *B*'s method of calling them exceptions. I believe that if we look at the supposed monsters and exceptions closely, we will find that they do in fact satisfy Euler's theorem.

T: Really…

A: What about my urchin (*A* points it out.) with its 12 star-pentagon faces?

R: I don't see 12 star-pentagons. I see 60 ordinary triangular faces, 90 edges, and 32 vertices. Hence V + F - E = 2 just as Euler predicted. There are no monsters, just monstrous interpretations. You just need to correctly recognize and interpret what you are seeing.

A: I've heard enough of this brainwashing. *T*, please show us your method.

T: Let *R* go on.

R: I have made my point.

O: I don't understand. Surely our goal is to find out exactly which polyhedra satisfy the condition that V + F - E = 2.

Scene 9: Generalizing the Problem

Z: No, our problem was just to find out what relationships exist between V, E, and F for polyhedra in general. The fact that we happened to have stumbled

on the Eulerian polyhedra first should not limit our study. We saw soon enough that there are at least as many non-Eulerian polyhedra as there are Eulerian. Why not look at when V + F − E = −6 or 28 or even 0? Why are they any less interesting?

S: You're right. We studied V + F − E = 2 because we thought it was true. Now that we know it isn't, we must look for a deeper more basic conjecture, one that will be true for all polyhedra.

O: Let's first solve the Euler problem before we move on. I want to understand exactly why some polyhedra are Eulerian before we look at more general questions. I want to find the secret of Eulerianess!

Z: I understand your resistance, O. You have fallen in love with the problem of finding out where God drew the boundary between Eulerian and non-Eulerian polyhedra. But how do you know there is such a line? Maybe Eulerianess is just an accidental property of some polyhedra with no great mystical ramifications. Perhaps Eulerianess is not part of some great order in the Universe.

S: Now we're really lost. With Eulerianess gone, what chance can we possibly have of finding any new order in the chaos of polyhedra and the relationships between vertices, edges, and faces?

B: We found the Eulerian pattern. Surely if we make an organized list of all the polyhedra we have found, we can find a new pattern and then work from there.

Scene 10: Induction versus Deduction

Z: Is that how you think mathematics is created? Just trying one guess after another hoping to stumble upon a pattern?

B: Yes, mathematical knowledge always starts with observation and some insightful discovery. Deductive reasoning only starts after the initial inductive phase.

S: Our first discovery of V + F - E = 2 was sheer luck. And still, it ended in a mess. It is even less likely that we'll come upon anything useful a second time.

B: How else can we start?

Z: I don't need any data to start. I have neither the time, money, nor interest to catalogue and categorize every last polyhedron and then test one formula after another.

B: What will you do then? Lie down on a couch, shut your eyes, and wait?

Z: Exactly, I must start with an idea.

B: And where will this idea come from?

Scene 11: First Principles

Z: The idea is already in our minds. It comes from the background knowledge that we already possess. In this case, we knew that for any polygon, V = E. A polygon is a system of polygons containing only one polygon. A polyhedron is a system of polygons containing more than one polygon. For a polyhedron, V does not equal E. So we need to look at why, when we go from one to more than one polygon, V suddenly stops equaling E.

S: So we start with E = V or E - V = 0. Adding a polygon, two edges become one and four vertices become two. So E goes down by one and V goes down by two causing E - V to go up by one. Hence now E - V = 1. (*S* demonstrates with model.) No matter how they go together, we will always loose one more vertex than edges, so E - V will always go up by 1. If E - V goes up by one, V - E goes down by one. Of course F will always go up by one. So V + F - E will stay the same. Thus V + F - E = 1 for a single polygon and it will remain one as we add faces.

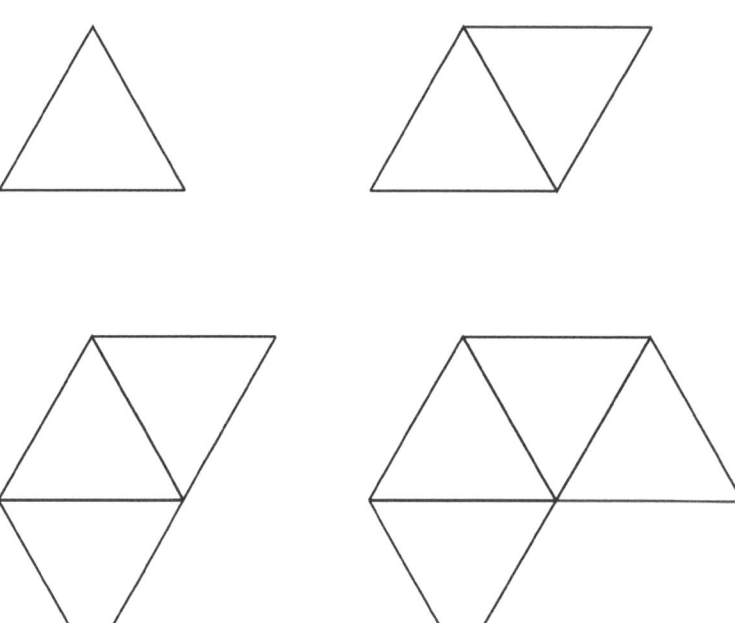

C: But polyhedra have V + F - E = 2.

S: Yes, but that is because my system only leads to open polyhedra with one open face. Once the polyhedron is closed with the final capping face, F increases by one making V + F - E = 2.

Z: So you see, I did not need to start with inductive reasoning.

B: I disagree. You merely pushed back the observation. Your starting point, that V = E for polygons, was certainly an inductive start. How, in fact, did you get V = E?

Z: I was deeply shocked when I realized that V - E = 0 for the triangle. I knew, of course, that for one edge, V - E = 1. I also knew that adding an edge increases both V and E by one. (Z shows with model.) Thus V - E remains equal to one. Then I realized that this method will only result in open systems of edges. When I close the system to make a polygon, I add one edge but no vertices and hence V - E decreases from one to zero.

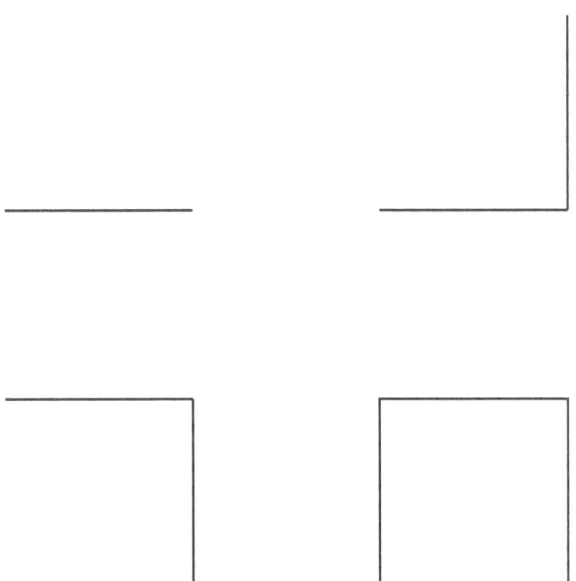

B: Now you've just pushed it back further. Didn't you inductively observe that V − E = 1 for an edge?

Z: Actually I started with V = 1 for a point.

B: So wasn't that an inductive beginning?

Z: I suppose if I said I started with empty space, you would accuse me of observing nothing.

T: We need to close this discussion for now.

S: We can't stop here. Nothing has been settled.

T: Mathematical inquiry begins and ends with questions.

B: But I didn't start with questions. Now I have nothing but questions.

Questions to Ponder

- How is Euler's theorem related to the existence of dual pairs?
- Is mathematics a creation of the human mind, or did it always exist?
- Does mathematics come from observation, inspiration, induction, deduction?
- Can anything be completely defined?
- Can a mathematical relationship be correct, even if it cannot be proved?
- Can a natural pattern or rule exist, even if there is no logical rule or mechanism to cause it?

Appendix

Lesson Plans

This appendix contains suggested lesson plans for teaching students about the Platonic solids. The lessons are general enough that they can be easily adapted to any age level from elementary school through college. The preceding chapters should serve well as a teaching resource for these lessons. I strongly suggest, however, that you also visit the companion website: *www.PlatonicSolids.info,* and that you also consider getting a copy of the video, *Platonic Solid Rock*, which is available on the website.

Lesson One — Building Paper Platonic Solids

Preparation

- Photocopy the patterns from chapter three for triangles, squares, and pentagons onto colored card stock.

- Score the fold lines with a straightedge and a compass point or with a dead ballpoint pen.

- Cut out the shapes.

- Using chapter one as a guide, build each Platonic solid. Experiment with different types of glue, or use a stapler, until you have a technique that works. You can either hide the tabs on the interior or have them on the outside.

- Once you have built all five solids, decide how much preparation work you want to do and how much the students will do in terms of scoring and cutting out the shapes.

- Note: If you find the paper shapes too difficult to use, you can substitute plastic snap-together pieces such as Polydron™.

Procedure

- Display your paper Platonic solids for the students to see.
- Provide students with construction materials.
- Guide students through the construction of their own Platonic solids.

Notes

_____

_____

_____

_____

_____

# Lesson Plans

Lesson Two — Deriving Euler's Formula

Preparation

- Create a handout with a grid such as the following:

Shape Name	Edges	Vertices	Faces
tetrahedron			
octahedron			
icosahedron			
cube			
dodecahedron			

- Create a poster-sized version of the same grid.

Procedure

- Pass out the shapes the students created in the previous lesson.

- Pass out the grids.

- Have students work in small groups counting edges, vertices, and faces and filling in their grids.

- When the groups have finished counting, create a class consensus of the counts and fill in the poster-sized grid.

- Ask students to try to find a pattern that holds for all of the shapes. You can model this by trying out a relationship such as Faces + Edges − Vertices.

- Hopefully, they will discover some variation of Euler's formula: $V + F - E = 2$. If students come up with different variations, discuss whether they are equivalent.

Lesson Plans

Notes

Lesson Three — Generalizing Euler's Formula

Preparation

- Create a handout with a grid such as the following:

Shape Name	Vertices (V)	Faces (F)	Edges (E)	V + F - E

- Create a poster-sized version of the same grid.

- Prepare an assortment of paper triangles, squares, pentagons, and hexagons from the patterns in chapter three.

Procedure

- Give each student a new grid.

- Give students access to a variety of shapes.

- Ask students to make up their own polyhedra. They can use whatever combination of face shapes they want. They should make up names for their new polyhedra.

- Have students count vertices, faces, and edges for their new polyhedra and fill in the chart.

- If all goes well, the last column will have all twos for entries.

- Have students transfer their data to the poster.

★ Discuss what this means about Euler's formula for polyhedra.

Notes

_____

_____

_____

_____

_____

Lesson Four — Proof?!

Preparation

- Read the play, *Proof?!,* in chapter five.

- Decide how much, if any of it, to share with your students.

- Make copies of the play for students. (There is a downloadable version on the website, *www.PlatonicSolids.info.*)

- If you are feeling particularly ambitious, create props.

Procedure

- Ask students to read the play through to themselves.

- Assign parts and read the play as a class.

- Follow with a discussion of Euler's formula and the play.

- Some other concepts you may want to cover are as follows:

  - Proof
  - Logic
  - Truth
  - Thought experiment
  - Definition
  - Undefined terms
  - Conjecture

- Theorem
- Lemma
- Counterexample
- Deduction
- Induction
- Argument
- Mathematics: invented or discovered?

✯ Consider giving a writing assignment about some major ideas in the play.

Notes

Lesson Five — Dual Pairs

Preparation

- Cut out paper squares, triangles, and pentagons photocopied from the patterns in chapter three.

- Cut out the three types of stellate patterns.

- Create the five Platonic solids.

- Use the stellate patterns to glue a small pyramid onto each face of each Platonic solid.

- You now have two versions of two of the compound polyhedra and one version of the third.

- Prepare enough paper patterns for all of the students to create their own compound polyhedra.

Procedure

- Tell the students how the Platonic solids are divided up into dual pairs and give them access to models of each solid.

- Ask the students if they see a pattern in the number of faces meeting at each vertex and the number of sides on each face of the members of the dual pairs.

- Also ask students if they see a pattern in the total number of vertices, the total number of faces, and the total number of edges of the members of the dual pairs.

- Revisit Euler's formula and see what happens to the result if the values for V and F are swapped.

- Show the students your compound polyhedra and ask them to try to visualize the component Platonic solids.

- Pass out the materials and help students make their own compound polyhedra.

Notes

Lesson Six — Research Reports

Preparation

- Create a list of possible research project questions.

- Here are some possible ideas:

    - What other groups of polyhedra exist?

    - What do you get when you connect the midpoints of the edges of the Platonic solids?

    - What do you get when you connect the points of the compound polyhedra?

    - What are the shapes of the different cross-sections of the Platonic solids?

    - What are the shapes of the different shadows of the Platonic solids?

    - How can you calculate the surface areas of the Platonic solids?

    - How can you calculate the volumes of the Platonic solids?

    - How do the surface areas of the Platonic solids compare to the surface area of a sphere?

    - How did Kepler use the Platonic solids?

    - What significance did Plato give to his solids?

    - Which polyhedra can be close packed?

    - What possible truss structures can be made from tetrahedra and octahedra?

    - How are geodesic domes related to polyhedra?

- What kinds of polyhedra are naturally formed as crystals?
- What polyhedra are found in molecular structures?
- What is Buckminsterfullerine (Bucky Balls)?
- What polyhedra can be found in the art of M. C. Escher?
- What is the golden mean and how is it related to the Platonic solids?
- Also consider the *Questions to Ponder* from the ends of chapters one, two, and five.

Procedure

- Assign students to do a research project.
- They can use the links on the *www.PlatonicSolids.info* site.
- They can search the web themselves.
- They can look for books and encyclopedias in the library.
- They can build models and conduct their own experiments.

Notes

_____

_____

_____

www.ingramcontent.com/pod-product-compliance
Lightning Source LLC
Chambersburg PA
CBHW051020180526
45172CB00002B/421